Innovations in Perovskite Solar Cell Materials and Devices - Cutting-Edge Research and Practical Applications

Edited by Rajendran Venkatachalam, Anita R. Warrier and Raja Mohan

Published in London, United Kingdom

Innovations in Perovskite Solar Cell Materials and Devices - Cutting-Edge Research and Practical Applications
http://dx.doi.org/10.5772/intechopen.1004511
Edited by Rajendran Venkatachalam, Anita R. Warrier and Raja Mohan

Contributors
Anjali Oudhia, Asha Chauhan, Ashok Kumar Shrivastav, Haribhahu Borate, Kiran Diwate, Lindiwe Khotseng, Lyle A. September, Nagaraj Gopinathan, Ntalane S. Seroka, Ntombizonke Kheswa, Rahul Ahir, Sadhu K. Kolekar, Sivasami Kolandasamy, Syed Abdul Gafoor Sathik Basha, Thangalakshmi Sivalingam, Tingting Yin

First published in London, United Kingdom, 2025 by IntechOpen
IntechOpen is the global imprint of INTECHOPEN LIMITED, registered in England and Wales, registration number: 11086078, 167-169 Great Portland Street, London, W1W 5PF, United Kingdom

For EU product safety concerns: IN TECH d.o.o., Prolaz Marije Krucifikse Kozulić 3, 51000 Rijeka, Croatia, info@intechopen.com or visit our website at intechopen.com.

British Library Cataloguing-in-Publication Data
A catalogue record for this book is available from the British Library

Innovations in Perovskite Solar Cell Materials and Devices - Cutting-Edge Research and Practical Applications
Edited by Rajendran Venkatachalam, Anita R. Warrier and Raja Mohan
p. cm.
Print ISBN 978-1-83769-884-4
Online ISBN 978-1-83769-883-7
eBook (PDF) ISBN 978-1-83769-885-1

If disposing of this product, please recycle the paper responsibly.

Meet the editors

Dr. Rajendran Venkatachalam is a distinguished expert in Nanotechnology and Biomaterials with over 30 years of academic experience. Ranked among the top 2% of scientists worldwide in Material Science by Elsevier BV and Stanford University, he has received over 18 prestigious fellowships, including DAAD-Germany and Indo-German exchanges. Dr. Rajendran has authored 265+ peer-reviewed papers, 125 conference papers, 26 textbooks, and 2 research books and holds 14 patents. He has visited 28 countries for academic collaborations and signed 25+ MoUs with international universities. In addition to his research, Dr. Rajendran serves as an editor for 14 journals and reviews 23 high-impact publications in Nanotechnology and Materials Science.

Dr. Anita R. Warrier is a Professor of Physics and the Dean of the Faculty of Science and Humanities at AMET University in Chennai, India. With a Ph.D. from Cochin University of Science and Technology in India, she is an expert in nanophotonics, photocatalysis, and sustainable energy technologies. A dynamic researcher, she has led several high-impact projects funded by the Department of Science and Technology (DST), the Council of Scientific and Industrial Research (CSIR), the University Grants Commission (UGC) and the Department of Atomic Energy (DAE), focusing on advanced nanomaterials for environmental remediation and energy storage. Dr. Warrier is also a prolific author and inventor, holding multiple patents and publishing extensively in top-tier journals. Her work on photothermal techniques for non-destructive evaluation has earned her recognition, including the prestigious DST-SERB Early Career Award. Passionate about mentoring, she has guided numerous Ph.D. students toward successful careers in science.

Dr. Raja Mohan is a distinguished academician and researcher with expertise in Materials Science, Chemical Engineering, and Nanotechnology. He holds a Master's degree in Materials Science and Technology from Madurai Kamaraj University in India and a Ph.D. in Chemical Engineering from Kyung Hee University in South Korea. Dr. Mohan has held faculty positions globally and published extensively in international journals. He has filed multiple patents in nanotechnology and led several funded research projects, including those from India's Department of Science and Technology (DST) and the National Semiconductor Hub in Saudi Arabia. His research focuses on smart polymers, hybrid nanocomposites, and energy storage applications. Dr. Mohan has delivered invited lectures at prestigious institutions and received multiple national awards for his contributions to research and academia

Contents

Preface

The quest for sustainable and efficient energy solutions has never been more urgent. As the global demand for renewable energy accelerates, the search for technologies that can efficiently harness solar power has become a central focus of scientific research and innovation. Among the most promising advancements in solar energy is the development of perovskite-based materials and devices, which have garnered significant attention for their remarkable efficiency, low production costs, and potential for large-scale deployment.

Innovations in Perovskite Solar Cell Materials and Devices – Cutting-Edge Research and Practical Applications explores this groundbreaking field, offering a comprehensive and up-to-date overview of the state-of-the-art advancements in perovskite solar cell technology. This book is designed for researchers, engineers, and industry professionals interested in understanding the complexities of perovskite solar cells, their materials, and the practical challenges and opportunities they present for real-world applications.

The book is structured around key areas of research and development in this rapidly evolving field. In the initial chapters, we explore the foundational principles of solar cell technology, providing a thorough understanding of the photovoltaic effect, the different types of solar cells, and their respective materials. This is followed by examining perovskite materials, which have revolutionized the solar energy landscape due to their unique properties, such as their high absorption coefficient, tunable bandgap, and ease of fabrication. We also explore novel approaches to synthesizing, characterizing, and optimizing perovskite films and devices.

The subsequent chapters focus on designing and fabricating perovskite solar cells, highlighting the cutting-edge research that has enabled significant advancements in power conversion efficiency. We cover the latest breakthroughs in material science, device engineering, manufacturing processes, and the challenges that remain in terms of stability, scalability, and commercial viability. Finally, we focus on the practical applications of perovskite solar cells, considering their potential integration into existing energy infrastructure, their role in advancing energy access, and the future prospects for commercialization.

Throughout the book, we strive to present a balanced view, combining theoretical insights with practical considerations. By showcasing the latest research alongside real-world applications, we aim to provide a resource that is not only informative but also inspiring for those working in the field of solar energy.

As the field of perovskite solar cells continues to evolve rapidly, we hope this book will serve as a valuable reference and catalyst for further innovation. By understanding this emerging technology's potential and challenges, we can move one step closer to achieving a sustainable and energy-efficient future.

Dr. Rajendran Venkatachalam
AMET University,
Kanathur, India

Dr. Anita R. Warrier
Department of Physics,
AMET University,
Kanathur, India

Dr. Raja Mohan
Department of Chemistry,
AMET University,
Kanathur, India

Chapter 1

Advanced Developments in Fabrication Techniques and Bandgap Engineering in Perovskite Solar Cells (PSC)

Lyle A. September, Ntombizonke Kheswa, Ntalane S. Seroka and Lindiwe Khotseng

Abstract

Perovskite materials have attracted attention because of their favorable features, which include configurable bandgap, improved charge transport, and unique electronic structure. The authors emphasize the most recent developments in PSC synthesis, innovative device architecture, bandgap engineering, charge transport processes, and stability. The need for a comprehensive design principle that surpasses the iterative process of the current synthesis of perovskite materials is highlighted. This is essential for developing high-quality, defect-free crystal structures with controlled dimensionality, and the choice of the deposition process is crucial to improve structural morphology in the production of PSC. This chapter also highlights that tandem and hybrid perovskite research is still in early development and emphasizes the need for basic research to fully understand the optical, electrical, and physical characteristics of the material before advances in industry.

Keywords: perovskite solar cells, two-step perovskite deposition, bandgap engineering, optoelectronic properties, material stability

1. Introduction

Finding safe, dependable, and clean energy sources is aided by the steady rise in technological innovation, global warming, and rising living standards. Because of this, conventional fossil fuel energy sources are insufficient to maintain both environmental preservation and the sustainable expansion of human society [1]. Consequently, there is a great deal of momentum behind rigorous scientific research to find sustainable renewable energy, with one of the most practical, dependable, and inexhaustible resources being solar energy. In general, a solar cell is an apparatus that converts light energy directly into electrical energy through photovoltaic effects or photochemical reactions. Perovskite solar cells have garnered a lot of attention lately due to their exceptional photovoltaic capabilities [2].

Optoelectronic device research has been thoroughly investigated in developing solar photovoltaic materials throughout the last few decades. The current rush in research and commercial attention is due to the critical role that perovskites play in the solar energy industry, with progressive properties in high power conversion efficiency (PCE) and low fabrication cost [3]. The rapid advance in efficiencies of perovskite/silicon (Si) tandem solar cells and metal halide semiconductor materials shows significant improvements in perovskite technology. Recent advances in metal halide semiconductors have been acknowledged as one of the most interesting possible light-absorbing materials with remarkable charge carrier transport performance in optoelectronic materials, particularly for photovoltaic applications. Optoelectronic devices based on metal halide semiconductors have increased developments in perovskite device performance [4, 5].

Perovskite solar cells (PSCs) have garnered a lot of attention from researchers lately because of their outstanding optical and electrical features, ease of fabrication, low production costs, and increased efficiency. As shown in **Figure 1**, recent studies have demonstrated an increase in efficiency and that monolithic perovskite/Si tandem cells can exceed the Shockley-Queisser limit for silicon, yielding a record of 33.7% created by KAUST. As single-junction silicon photovoltaics (PVs), which currently dominate the industry, get closer to their theoretical efficiency limit of 29%. Hybrid organic–inorganic perovskites (HOIPs) are one of the most well-liked next-generation solar materials due to their remarkable certified power conversion efficiency (PCE), which exceeds 25%. The National Renewable Energy Laboratory (NREL) has confirmed that Longi Green Energy, a Chinese photovoltaic manufacturer, achieves maximum efficiency. Considering the remarkable power conversion efficiency that has been achieved thus far, the main barriers to PSC commercialization are material selection, the fabrication processes used, band gap engineering, degradation, hysteresis effect, and stability problems [6]. These notable challenges are motivating academics to create efficient PSCs.

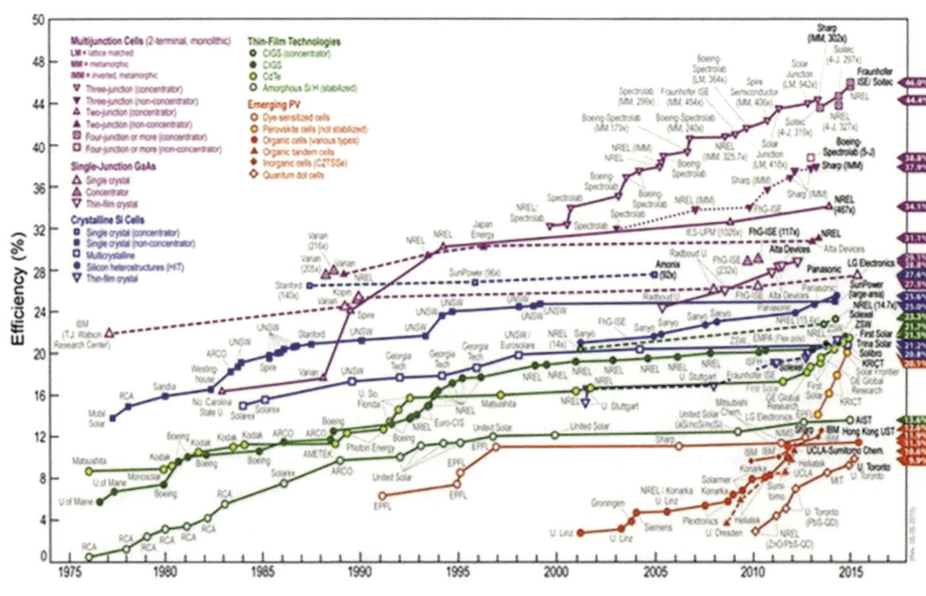

Figure 1.
Best solar cell efficiencies reported by the National Renewable Energy Laboratory (NREL).

2. Recent advances

2.1 Background

PSCs generally have the formula ABX_3, where X is a halogen anion, B is a divalent cation (Pb^{2+} or Sn^{2+}), and A is a monovalent organic or inorganic cation, most often methylammonium (MA+), potassium (K), or cesium (Cs), illustrated in **Figure 2** [7]. The architecture of PSCs is also essential to their effectiveness because carrier generation and extraction depend heavily on the morphology of the various layers and their interfacial characteristics. PSCs are typically created by stacking perovskite active layers, anode/cathode, and carrier transport layers (also known as electron/hole transport layers, or ETLs/HTLs, respectively) to create a multi-interface system [8]. Based on a unique crystal structure called perovskite, perovskite solar cells absorb light by employing substances that are hybrid organic–inorganic, such as tin halide or lead halide. Electrons in the perovskite layer are excited by sunlight, which results in the creation of electron–hole pairs, or excitons. At the interfaces where the perovskite layer joins adjacent charge transport layers—such as titanium dioxide for electrons and spiro-OMeTAD for holes—efficient charge separation takes place [9]. Electrodes at the cell's contacts then gather electrons and holes, which when they travel *via* an external circuit, produce an electric current. Electrical power can be produced from this current.

The unique characteristics of perovskite materials, such as their high carrier diffusion length (greater than 1 μm), high electron mobility (800 cm^2/Vs), and electrical characteristics (charge transport behavior), make them one of the best materials for the next generation of photovoltaic technology [10]. A multitude of different components can be combined to generate perovskite structures. Its features can be altered by modifying its composition and structure. To assess the stability of the perovskite structure, the parameters "tolerance factor" (t) and "octahedral factor" (u) should be considered during the design process. The perovskite's t and u range from 0.78 to 1.05 and 0.44 to 0.9, respectively, when it is in a stable structure [11].

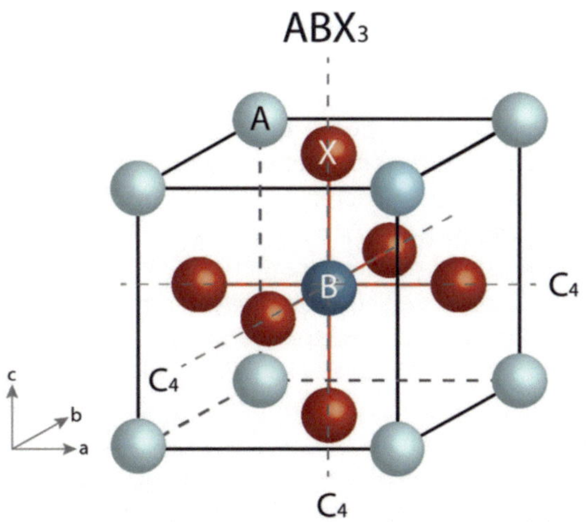

Figure 2.
The perovskite crystal structure with the molecular formula of ABX_3 (A: Cs+, Rb+, MA+ = $(CH_3NH_3)^+$, or FA^+ = $(CH(NH_2)_2)^+$; B: Pb^2+ or Sn^{2+}; and X: Cl−, Br−, or I−) Liao et al. [7].

The lack of stability studies in current perovskite-based cell research trends limits its applicability in outdoor applications. There would be very few commercial uses for perovskite solar cells if research on their stability and rate of degradation was not conducted. According to recent research on PSCs, chemical stability is greatly influenced by oxygen, temperature, moisture, UV light, and solution processing techniques [12]. PbI_2 and methylammonium iodide (MAI), for example, react very easily at low temperatures when driven by hydrogen and ionic interactions, making the synthesis of perovskite surprisingly straightforward experimentally [13].

Perovskite solar cells (PSCs) have advanced rapidly, with remarkable performance uptake from 3.8% to >26%; yet, the main obstacles to their commercialization are their low stability and toxicity. Many fabrication techniques and changes have been investigated to date in order to enable its commercialization [14]. The excellent light absorption across a wide spectrum, a possibility for low-cost manufacture by solution processing, and the theoretically high efficiency of perovskite solar cells make them an attractive technology [15]. The goal of ongoing research is to make perovskite solar cells a viable commercial alternative to silicon solar cells by improving stability and addressing environmental issues like lead toxicity in specific compositions [16].

2.2 Materials

Perovskite materials have attracted a lot of attention due to their unique optical, thermal, and electromagnetic capabilities, as well as their cubic lattice-nested octahedral structures. Perovskite materials have the following four characteristics: these materials have strong optical absorption coefficients, good photoelectric properties, and low exciton binding energies [17]. Perovskites have a high dielectric constant that facilitates efficient electron and hole photogeneration; furthermore, they can absorb solar energy effectively while successfully transporting electrons and holes from 100 nm to more than 1 μm. When employed in solar photovoltaic applications, these perovskite materials provide noticeably greater open-circuit voltages and short-circuit current densities [18]. Perovskite materials can also be classified as lead-free perovskites, all-inorganic perovskites, and organic–inorganic hybrid perovskites based on their chemical composition.

To further understand PSC performance and potential, a few perovskite materials are described. **Figure 3** describes the charge transfer and energy levels in flexible perovskite solar cells (F-PSC), because of a lower exciton binding energy, the photogenerated electron–hole pairs in the perovskite layer dissociate readily to free carriers [20]. Furthermore, the transparent conductive plastic substrates are coated with the ETL, which functions as a charge-selective layer. A perovskite absorption layer is then created using a spin-coating process. After that, the back electrodes and the HTL are deposited to create a normal n-i-p structure planar F-PSC. Additionally, the charge transport layers' locations are reversed for an inverted p-i-n planar F-PSC [19].

Many materials have been developed for use as ETL, HTL, and active layers in PSCs, including SiO_2, TiO_2, CdS, ZnO, PEDOT:PSS, NiO_x, Spiro-OMeTAD, PCBM, $MAPbI_3$, $MAPbBr_3$, $MASnI_3$, $CsPbBr_3$, $FASnI_3$, and so on [21]. Specifically, Spiro-OMeTAD, the hole-transporting material used in perovskite solar cells, is more expensive and needs a labour-intensive production procedure [22]. New low-cost hole transport materials must be developed in order to support the commercialization of perovskite solar cells. The methylammonium lead iodide ($MAPbI_3$—$CH_3NH_3PbI_3$) perovskite solar cell's favorable optical and electrical absorption coefficients contribute to its exceptional performance. Furthermore, the $MAPbI_3$ absorber layer has one

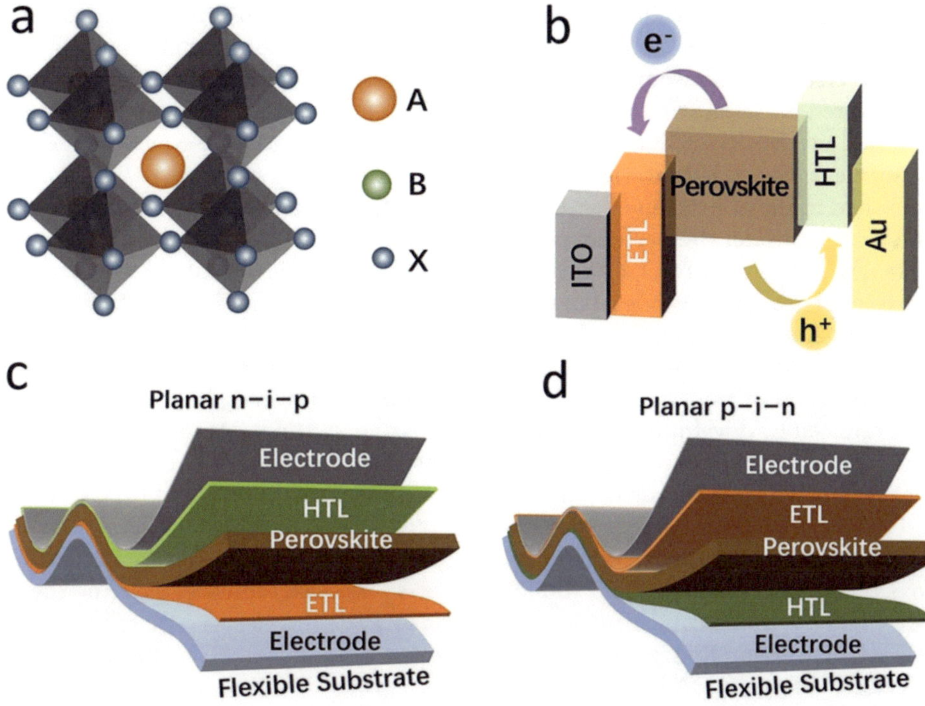

Figure 3.
(a) Diagram showing the typical crystalline structure of ABX3 perovskite. (b) Diagram showing the relative energy levels and charge transmission. Typical flexible PSC devices with p-i-n and n-i-p architectures, (c) and (d), respectively [19].

of the highest open-circuit voltages (V_{oc}), the lowest V_{oc} loss/deficit, and the lowest exciton binding energy [23]. These characteristics lead to improved charge transport, respectable charge carrier mobilities, and extended charge carrier diffusion lengths, which make it a viable option for photovoltaic applications. The $MAPbI_3$ material-based PSC has a very good PCE of about 23.7%, an acceptable band gap of 1.55 eV, remarkable electron mobility of 24.0 ± 7 $cm^2\ v^{-1}\ s^{-1}$, and hole mobility of 105 ± 35 $cm^2\ v^{-1}s^{-1}$ [24]. The most extensively studied perovskite solar cell is $MAPbI_3$-based, and it has a wide range of possible applications.

Similarly, other materials such as $FAPbI_3$ and $CsPbI_3$ are utilized. The 1.43 eV narrow bandgap of $FAPbI_3$ is within the optimal range for achieving very high single-junction solar cell efficiency. Furthermore, compared to $MAPbI_3$, $FAPbI_3$ has superior semiconducting characteristics, such as a lower electron-photon coupling, a longer carrier lifetime, and a smaller charge carrier effective mass. Furthermore, the MA escape-induced device deterioration of $MAPbI_3$ results in low thermal stability. Additionally, formamidinium lead iodide ($FAPbI_3$)-based perovskite solar cells have a 19.09% efficiency at room temperature. The $FAPbI_3$ material has a narrow band gap, a high light-absorption coefficient, and excellent photo- and thermal-stability [25]. However, PSCs will find the small band gap of $FAPbI_3$ perovskites to be very beneficial, which may speed up the commercialization of this fascinating perovskite material.

Furthermore, the industrial advancement and growth of PSCs will be hindered by the exceedingly hazardous nature of the Pb element utilized in perovskite solar cells. Since Pb is one of the toxic lead components in the $MAPbI_3$, replacing Pb with safer elements such as tin and germanium is the usual technique for eliminating toxic lead elements. Consequently, a low-toxicity or nontoxic component needs to be discovered in the future to replace Pb [26]. However, compared to PSCs using lead-based perovskites, PSCs using these alternative metal-based perovskites perform worse. This is due to the fact that the deterioration of perovskite devices has been linked to the migration of mobile ions [27]. As an example, there are two main problems that hinder the performance of PSC based on $CsPbI_3$. The low tolerance factor causes the preferred phase transition from the perovskite to the non-perovskite phase. Energy loss from energy-level imbalances and inevitable failures is the second, and it limits PCE [28].

Despite the superior performance of lead-based PSCs, lead toxicity necessitates the development of lead-free PSCs, shown in **Figure 4**. However, Pb exposure can be reduced by developing effective encapsulating methods. At the end of the module's lifecycle, it is imperative to ensure that all materials have been effectively recycled [29]. The large-scale production and commercialization of lead-based PSCs could release lead emissions into the environment, which could have a detrimental effect on the environment. Using a two-step spin coating technique, a low-toxicity perovskite material with 50% Pb and 50% Sn was produced while considering the toxicity and environmental effects of Pb [30]. The PSCs display an efficiency of 13.6% while utilizing Rs 5.3 Ω cm^2 and the device configuration ITO/PEDOT: PSS/ $CH_3NH_3Sn_{0.5}Pb_{0.5}I_3/C_{60}/Ag$ [22].

With a PCE of 29.74% and an ideal band gap of 1.3 eV, the FTO/TiO_2/$CH_3NH_3SnI_3$/ CuSbS configuration, a tin (Sn) halide perovskite device, exhibits improved absorption in the visible spectrum area [32]. Nevertheless, unlike lead perovskite, tin halide perovskite readily oxidizes to the oxidation state Sn4+, resulting in metallic properties and somewhat limited air stability. Similarly, germanium iodide perovskite ($MAGeI_3$), a distinct group 14 metal with significant promise for use in solar applications, is the most obvious substitute for lead in monovalent applications. The low performance

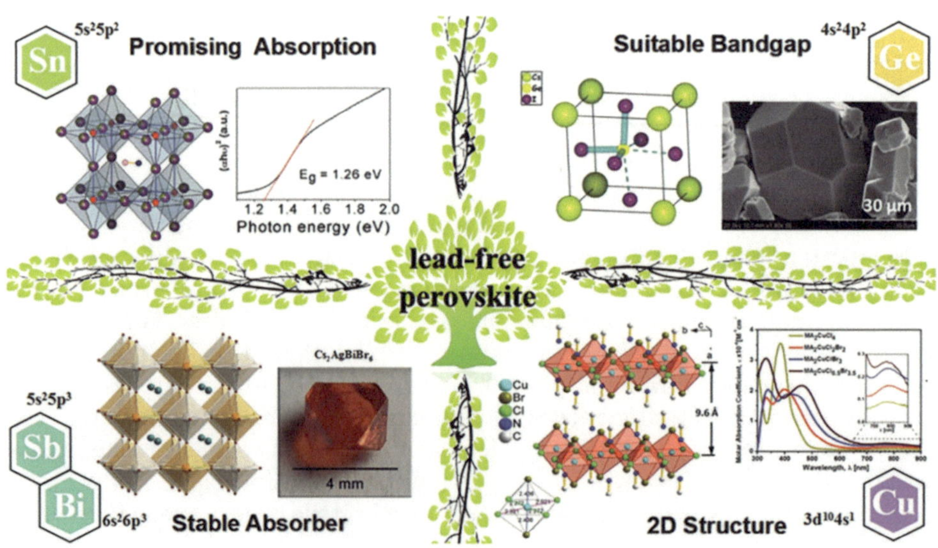

Figure 4.
Several common organic–inorganic hybrid lead-free perovskite structure types [31].

of the cell is caused by the oxidation-induced generation of Ge4+ ions, even though $MAGeI_3$ has a band gap of 1.9 to 2 eV [33].

Researchers are interested in enhancing photovoltaic cells (PCE) with lead-free halide perovskite materials because of their high carrier mobility, electron transfer kinetics, and environmental friendliness. The right band gap, low background charge carrier density, low carrier effective mass, high absorption coefficient, and greater stability at the ambient atmosphere are essential factors for the champion device [34]. Up to now, tin-based halide perovskites have demonstrated a 14% power conversion efficiency in a single junction. Compared to single junction, hybrid tin and germanium halide perovskite exhibits higher power conversion efficiency and improved stability in ambient air [35]. Although the morphology of bismuth halide perovskite material is poor due to rapid crystallization, it has superior stability, which could affect device performance [36].

Numerous tactics, such as the addition of additives, reducing agents, 2D structure components, and solvent engineering device structure, are used to get past these obstacles. The shape of the film is primarily responsible for the lead-free halide perovskite's high power conversion efficiency. The choice of solvent, concentration of the solution, sample annealing, and technique (one-step spin coating, two-step spin coating, vapor-aided solution process, antisolvent processing) all have an impact on the film morphology. When it comes to the device's performance, the stability of the perovskite material has finally proven to be a major challenge. Halide perovskite materials based on tin and germanium quickly oxidize because of chemical instability in the surrounding atmosphere. Although bismuth-based halide perovskite material has a high stability compared to other perovskites, its rapid crystallization results in poor morphology. Thus, in order to build a lead-free halide perovskite solar cell with improved stability and high efficiency, these aspects must be taken into consideration.

These limitations can be solved thanks to the unique advantages of the newly developed solar cells based on hybrid organic–inorganic perovskites. The novel class of optoelectronic semiconductors referred to as organic–inorganic hybrid metal halide PSCs has garnered significant attention from photovoltaic researchers across the world. The advantages of hybrid organic–inorganic perovskites, which are the basis of the recently released solar cells, allow for the circumvention of these constraints [37]. An improved nanocomposite's mechanical properties result from the addition of additives near the interface. Their continually rising productivity, easy-to-use solution processing methods, usage of reasonably priced, lightweight materials, and other notable attributes are the reasons for this [22]. Hybrid perovskite research is still in its early stages; therefore, before PSCs can be used in practical applications, more fundamental research is needed to fully comprehend their electrical, optical, and physical characteristics.

2.3 Fabrications

In recent years, since PSCs' performance has greatly increased, silicon solar cells are no longer thought to be a better option. In this work, we have discussed a number of ways to improve the efficiency of perovskite solar cells, such as using different manufacturing processes, creating new materials for perovskite and charge transport, utilizing band gap engineering developments recently, and addressing stability concerns [38]. Despite much research, there are still numerous barriers to PSC improvement.

Moreover, most techniques for synthesizing perovskite materials, such as solvent diffusion, hot injection, inverse temperature, solvent evaporation crystallization, and anti-solvent vapor assisted, are based on the solution process [39]. One-step and two-step deposition techniques are among the solution synthesis methods for perovskite film, which combine all these constituents in an ordered network structure during manufacturing. With the sizes of the B- and X-site ions adhering to the octahedral factor limit, the $[BX_6]^{4-}$ octahedra composed of a core metal cation (B^{2+}) and halogen anions (X^-) create tetragonal or cubic network architectures *via* corner-sharing connections. Diverse perovskite formation processes result from variations in the unit cell's composition and characteristics, which in turn provide a range of formation energies [40]. The choice and optimization of a particular deposition technique are influenced by a number of variables, including the highlighted application's specific requirements, cost-effectiveness, and scalability.

Therefore, creating more synthetic paths to the solution process is crucial. To further optimize the physical properties of perovskite materials, morphologically low-dimensional structures, including nanorods, nanocrystals, nanoplatelets, nanosheets, and nanowires, can be synthesized [40, 41]. The process of trial and error is still largely used in the synthesis of novel perovskite materials. The cultivation of superior crystal structures with controlled dimensionality with no flaws requires the creation of a thorough design principle [42].

In recent years, there has been a growing interest in perovskite-based double junction tandem solar cells, such as perovskite/silicon, perovskite/perovskite, and perovskite/$Cu(In, Ga)Se_2$ [43], although the one-step deposition approach has been used to create the majority of wide-band-gap perovskite solar cells to date. Researchers have successfully built a perovskite solar cell with an active layer of 1.63 eV bandgap and a PCE of 20.35% utilizing the two-step sequential deposition approach [44]. By using the same material method as in the one-step method, the compositional optimization allowed for an improvement in FF to 81.53% and PCE to 20.35%, which was thought to be the highest efficiency of wide-band-gap fabricated by the sequential deposition method, when compared to the best device (an FF of 72.66% and a PCE of 17.82%) produced by the two-step method [45]. Wide bandgap perovskite materials, with their appropriate bandgap range of 1.60–1.72 eV, are good absorber materials to achieve high efficiency and current alignment.

Illustrated in **Figure 5**, the structure of a tandem solar cell consists of an n-type silicon bottom sub-cell and a lead-free perovskite top cell (hole transport layer: NiO; absorber: $Cs_2AgBi_{0.75}Sb_{0.25}Br_6$; electron transport layers: PCBM and SnO_2). A substantial open-circuit voltage of 1.76 V, a short-circuit current of 16.01 $mA.cm^{-2}$, and a PCE of 24.4% are attained by the improved tandem solar cell [46]. The benefits of the lead-free perovskite tandem solar cell, which combines acceptable thermal expansion behavior with electrical parameter values and efficiency that are very competitive with standard silicon cells, are also demonstrated by the results of a simulation of the effect of the working temperature. This kind of lead-free perovskite/silicon tandem solar cell is an excellent candidate for possible uses in optoelectronics and photovoltaics due to its non-toxicity and remarkable moisture stability. Because of this, perovskites are especially well suited for tandem use. In the world of photovoltaic energy, tandem solar cells that mix silicon and perovskites have been deemed the best options [47].

One of the simplest ways to reduce the levelized cost of electricity (LCOE) is to increase power conversion efficiency through the use of the tandem strategy. The sensitized structure, or tandem film structure, of each of these perovskite solar cells

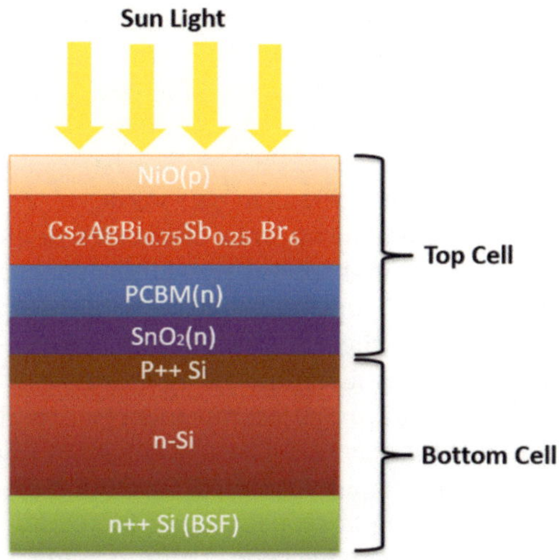

Figure 5.
Structure of a tandem solar cell consisting of an n-type silicon bottom sub-cell and a lead-free perovskite top cell [46].

involves the formation of heterojunctions between the perovskite material and a variety of other materials. Typically, this approach entails stacking two sub-cells on top of one another and joining them using a 2-terminal (2 T) monolithic design or a 4-terminal (4 T) architecture [48]. The tandem concept has drawn a lot of interest because of its high open-circuit voltage (VOC) and overall power conversion efficiency. The most efficient tandem solar cells are built on the foundation of III-V semiconductor technology. They have achieved an efficiency of 32.8% for two-junction cells and 38.8% for five-junction cells in non-concentrated AM1.5 sunlight [49].

Perovskite/silicon tandem solar cells have advanced quickly as a result of advances in interface engineering, tandem structure design, and perovskite component optimization. Furthermore, it is essential to consider important elements while developing 2 T perovskite/silicon tandem devices, such as the top perovskite cell with a broad bandgap [49]. A significant barrier to the development of low-cost, high-efficiency tandem devices has been the lack of a high-efficiency top cell with a bandgap larger than 1.5 eV. Despite this, the high cost of materials and epitaxial layer development has limited the application of these tandem devices to specific industries like concentrator photovoltaics.

3. Conclusion

This review chapter highlights the necessity of a thorough design production that goes beyond the process of trial and error to direct the synthesis of perovskite materials. Additionally, it recognizes the early stage of hybrid perovskite research, highlighting the need for basic study to fully comprehend the material's electrical, optical, and physical characteristics prior to any practical uses, especially in perovskite solar cells (PSCs). The investigation of substitute synthetic paths for the widely employed solution technique is emphasized, recognizing the drawbacks and looking for creative solutions.

A variety of materials are introduced for usage as active layers, hole transport layers (HTL), and electron transport layers (ETL) as this chapter explores the structure and tandem film structure of perovskite solar cells. Notably, the emphasis is on how important it is to create new, affordable hole transport materials in order to support perovskite solar cells. It is vital to create innovative, affordable hole transport materials to support the commercialization of perovskite solar cells. Furthermore, the industrial promotion and development of PSCs will be hampered by the exceedingly hazardous nature of the Pb element utilized in perovskite solar cells. Consequently, a low-toxicity or nontoxic component needs to be discovered in the future to replace Pb.

Furthermore, in terms of potential and practical industrial use, the monolithic two-terminal tandem cell is the best option. Si bottom cells that are sold commercially have an efficiency that approaches their theoretical maximum rather closely. The development of tandem devices is still hampered by a number of issues, such as scaling up manufacturing, the cost of acquisition, the inherent instability of perovskite materials, and performance improvement. The advantages of multifunctional cells have been highlighted due to the demonstration of intrinsic defects in single-junction solar cells. Due to the strengthening of tandem structures combined with perovskite/contact materials, efficiency records have been broken quickly. For the commercialization of perovskite/Si tandem technology, lab-scale 2 T perovskite/c-Si tandem efficiency has promise.

Author details

Lyle A. September[1*], Ntombizonke Kheswa[2], Ntalane S. Seroka[1,3*] and Lindiwe Khotseng[1]

1 Department of Chemistry, University of the Western Cape, Bellville, South Africa

2 NRF/iThemba LABS (Laboratory for Accelerator Based Sciences), Faure, South Africa

3 Energy Centre, Smart Places Cluster, Council for Scientific and Industrial Research (CSIR), Pretoria, South Africa

*Address all correspondence to: 3727392@myuwc.ac.za and 3754640@myuwc.ac.za

References

[1] Zhang J. Energy access challenge and the role of fossil fuels in meeting electricity demand: Promoting renewable energy capacity for sustainable development. Geoscience Frontiers. 2024;**15**(5):101873. DOI: 10.1016/j.gsf.2024.101873

[2] Giannouli M. Current status of emerging PV technologies: A comparative study of dye-sensitized, organic, and perovskite solar cells. International Journal of Photoenergy. 2021;**2021**(1):6692858. DOI: 10.1155/2021/6692858

[3] Chi W, Banerjee SK. Progress in materials development for the rapid efficiency advancement of perovskite solar cells. Small. 2020;**16**(28):1907531. DOI: 10.1002/smll.201907531

[4] Peng S, Yang Z, Sun M, Yu L, Li Y. Stabilizing metal halide perovskites for solar fuel production: Challenges solutions, and future prospects. Advanced Materials. 2023;**35**(45):2304711. DOI: 10.1002/adma.202304711

[5] Wang Y, Liu J, Liu Y, Li S, Xu X, Lou Z. Recent advances of Lead-free halide perovskites: From synthesis to applications. Journal of Materials Chemistry C. 2024;**12**:10267-10329. DOI: 10.1039/D4TC01556H

[6] Teixeira CO, Castro D, Andrade L, Mendes A. Selection of the ultimate perovskite solar cell materials and fabrication processes towards its industrialization: A review. Energy Science & Engineering. 2022;**10**(4):1478-1525. DOI: 10.1002/ese3.1084

[7] Liao C-H, Mahmud MA, Ho-Baillie AWY. Recent progress in layered metal halide perovskites for solar cells, photodetectors, and field-effect transistors. Nanoscale. 2023;**15**(9):4219-4235. DOI: 10.1039/D2NR06496K

[8] Ma Y, Gong J, Zeng P, Liu M. Recent progress in interfacial dipole engineering for perovskite solar cells. Nanomicro Letters. 2023;**15**(1):173. DOI: 10.1007/s40820-023-01131-4

[9] Shao S, Loi MA. The role of the interfaces in perovskite solar cells. Advanced Materials Interfaces. 2020;7(1):1901469. DOI: 10.1002/admi.201901469

[10] Roy P, Ghosh A, Barclay F, Khare A, Cuce E. Perovskite solar cells: A review of the recent advances. Coatings. 2022;**12**(8):1089. DOI: 10.3390/coatings12081089

[11] Liu B, Wang Y, Wu Y, Dong B, Song H. Novel broad spectral response perovskite solar cells: A review of the current status and advanced strategies for breaking the theoretical limit efficiency. Journal of Materials Science and Technology. 2023;**140**:33-57. DOI: 10.1016/j.jmst.2022.06.055

[12] Nair S, Patel SB, Gohel JV. Recent trends in efficiency-stability improvement in perovskite solar cells. Materials Today Energy. 2020;**17**:100449. DOI: 10.1016/j.mtener.2020.100449

[13] Jo Y-R, Tersoff J, Kim M-W, Kim J, Kim B-J. Reversible decomposition of single-crystal methylammonium lead iodide perovskite nanorods. ACS Central Science. 2020;**6**(6):959-968. DOI: 10.1021/acscentsci.0c00385

[14] Srivastava A et al. Lead metal halide perovskite solar cells: Fabrication,

advancement strategies, alternatives, and future perspectives. Materials Today Communications. 2023;**35**:105686. DOI: 10.1016/j.mtcomm.2023.105686

[15] Sharma R, Sharma A, Agarwal S, Dhaka MS. Stability and efficiency issues, solutions and advancements in perovskite solar cells: A review. Solar Energy. 2022;**244**:516-535. DOI: 10.1016/j.solener.2022.08.001

[16] Urbina A. The balance between efficiency, stability and environmental impacts in perovskite solar cells: A review. Journal of Physics: Energy. 2020;**2**(2):22001. DOI: 10.1088/2515-7655/ab5eee

[17] Han Y, Yue S, Cui B-B. Low-dimensional metal halide perovskite crystal materials: Structure strategies and luminescence applications. Advanced Science. 2021;**8**(15):2004805. DOI: 10.1002/advs.202004805

[18] Dey K, Roose B, Stranks SD. Optoelectronic properties of low-bandgap halide perovskites for solar cell applications. Advanced Materials. 2021;**33**(40):2102300. DOI: 10.1002/adma.202102300

[19] Li X et al. Progress and challenges toward effective flexible perovskite solar cells. Nanomicro Letters. 2023;**15**:206. DOI: 10.1007/s40820-023-01165-8

[20] Liu C, Chen J, Gao P. Enhanced perovskite photovoltaics with non-conjugated polymers: Recent developments and future prospects. ACS Applied Polymer Materials. 2024;**6**(7):3573-3610. DOI: 10.1021/acsapm.3c03111

[21] Suresh Kumar N, Naidu KCB. A review on perovskite solar cells (PSCs), materials and applications. Journal of Materiomics. 2021;**7**(5):940-956. DOI: 10.1016/j.jmat.2021.04.002

[22] Elangovan NK, Kannadasan R, Beenarani BB, Alsharif MH, Kim M-K, Hasan Inamul Z. Recent developments in perovskite materials, fabrication techniques, band gap engineering, and the stability of perovskite solar cells. Energy Reports. 2024;**11**:1171-1190. DOI: 10.1016/j.egyr.2023.12.068

[23] Mangrulkar M, Stevenson KJ. The Progress of additive engineering for $CH_3NH_3PbI_3$ photo-active layer in the context of perovskite solar cells. Crystals (Basel). 2021;**11**(7):814. DOI: 10.3390/cryst11070814

[24] Shukla A, Sharma VK, Gupta SK, Verma AS. Investigations of fundamental physical and thermoelectric properties of methylammonium lead iodide ($CH_3NH_3PbI_3$) perovskites. Materials Research Express. 2020;**6**(12):126323. DOI: 10.1088/2053-1591/ab619b

[25] Zhu J et al. A donor–acceptor-type hole-selective contact reducing non-radiative recombination losses in both subcells towards efficient all-perovskite tandems. Nature Energy. 2023;**8**(7):714-724. DOI: 10.1038/s41560-023-01274-z

[26] Podapangi SK et al. Green solvents, materials, and lead-free semiconductors for sustainable fabrication of perovskite solar cells. RSC Advances. 2023;**13**(27):18165-18206. DOI: 10.1039/D3RA01692G

[27] Xu Z et al. Functional layers of inverted flexible perovskite solar cells and effective Technologies for Device Commercialization. Small Structures. 2023;**4**(5):2200338. DOI: 10.1002/sstr.202200338

[28] Huang J et al. Advances in crystallization regulation and defect

suppression strategies for all-inorganic $CsPbX_3$ perovskite solar sells. Progress in Materials Science. 2024;**141**:101223. DOI: 10.1016/j.pmatsci.2023.101223

[29] Chen C-H, Cheng S-N, Cheng L, Wang Z-K, Liao L-S. Toxicity leakage, and recycling of Lead in perovskite photovoltaics. Advanced Energy Materials. 2023;**13**(14):2204144. DOI: 10.1002/aenm.202204144

[30] Lyu M, Yun J-H, Chen P, Hao M, Wang L. Addressing toxicity of Lead: Progress and applications of low-toxic metal halide perovskites and their derivatives. Advanced Energy Materials. 2017;**7**(15):1602512. DOI: 10.1002/aenm.201602512

[31] Liu Y. The development of low toxic and high efficient solar cells. Journal of Physics Conference Series. 2020;**1653**(1):12002. DOI: 10.1088/1742-6596/1653/1/012002

[32] Valeti NJ, Prakash K, Singha MK. Numerical simulation and optimization of lead free CH3NH3SnI3 perovskite solar cell with CuSbS2 as HTL using SCAPS 1D. Results in Optics. 2023;**12**:100440. DOI: 10.1016/j.rio.2023.100440

[33] Mehrabian M, Akhavan O, Rabiee N, Afshar EN, Zare EN. Lead-free MAGeI3 as a suitable alternative for MAPbI3 in nanostructured perovskite solar cells: A simulation study. Environmental Science and Pollution Research. 2023;**30**(19):57032-57040. DOI: 10.1007/s11356-023-26497-1

[34] Farooq U et al. Bandgap engineering of lead-free ternary halide perovskites for photovoltaics and beyond: Recent progress and future prospects. Nano Energy. 2022;**92**:106710. DOI: 10.1016/j.nanoen.2021.106710

[35] Pitaro M, Tekelenburg EK, Shao S, Loi MA. Tin halide perovskites: From fundamental properties to solar cells. Advanced Materials. 2022;**34**(1):2105844. DOI: 10.1002/adma.202105844

[36] Tailor NK, Satapathi S. Structural disorder and spin dynamics study in Millimeter-sized all-inorganic Lead-free Cesium bismuth halide perovskite single crystals. ACS Applied Energy Materials. 2020;**3**(12):11732-11740. DOI: 10.1021/acsaem.0c01849

[37] Krishnamurthy S et al. Organic–inorganic hybrid and inorganic halide perovskites: Structural and chemical engineering, interfaces and optoelectronic properties. Journal of Physics D: Applied Physics. 2021;**54**(13):133002. DOI: 10.1088/1361-6463/abd0ad

[38] Miah MH, Khandaker MU, Rahman MB, Nur-E-Alam M, Islam MA. Band gap tuning of perovskite solar cells for enhancing the efficiency and stability: Issues and prospects. RSC Advances. 2024;**14**(23):15876-15906. DOI: 10.1039/D4RA01640H

[39] Ke L, Luo S, Ren X, Yuan Y. Factors influencing the nucleation and crystal growth of solution-processed organic lead halide perovskites: A review. Journal of Physics D: Applied Physics. 2021;**54**(16):163001. DOI: 10.1088/1361-6463/abd728

[40] Zhang A, Lv Q. Organic-inorganic hybrid perovskite nanomaterials: Synthesis and application. Chemistry Select. 2020;**5**(41):12641-12659. DOI: 10.1002/slct.202003659

[41] Chenna P, Gandi S, Pookatt S, Parne SR. Perovskite white light emitting diodes: A review. Materials Today Electronics. 2023;**5**:100057. DOI: 10.1016/j.mtelec.2023.100057

[42] Lou Y et al. Perovskite single crystals: Dimensional control, optoelectronic properties, and applications. Materials

Today. 2023;**62**:225-250. DOI: 10.1016/j.mattod.2022.11.009

[43] Ašmontas S, Mujahid M. Recent Progress in perovskite tandem solar cells. Nanomaterials. 2023;**13**(12):1886. DOI: 10.3390/nano13121886

[44] Gholipour S, Saliba M. Chapter 1 - bandgap tuning and compositional exchange for lead halide perovskite materials. In: Pazoki M, Hagfeldt A, Edvinsson T, editors. Characterization Techniques for Perovskite Solar Cell Materials. Micro and Nano Technologies. Amsterdam: Elsevier; 2020. pp. 1-22. DOI: 10.1016/B978-0-12-814727-6.00001-3

[45] Liu X et al. Highly efficient wide-band-gap perovskite solar cells fabricated by sequential deposition method. Nano Energy. 2021;**86**:106114. DOI: 10.1016/j.nanoen.2021.106114

[46] Amri K, Belghouthi R, Aillerie M, Gharbi R. Device optimization of a Lead-free perovskite/silicon tandem solar cell with 24.4% power conversion efficiency. Energies (Basel). 2021;**14**(12):3383. DOI: 10.3390/en14123383

[47] Chen Q et al. Recent Progress of wide bandgap perovskites towards two-terminal perovskite/silicon tandem solar cells. Nanomaterials. 2024;**14**(2):202. DOI: 10.3390/nano14020202

[48] Khan F, Rezgui BD, Khan MT, Al-Sulaiman F. Perovskite-based tandem solar cells: Device architecture, stability, and economic perspectives. Renewable and Sustainable Energy Reviews. 2022;**165**:112553. DOI: 10.1016/j.rser.2022.112553

[49] Torabi N et al. Progress and challenges in perovskite photovoltaics from single- to multi-junction cells. Materials Today Energy. 2019;**12**:70-94. DOI: 10.1016/j.mtener.2018.12.009

Chapter 2

High-Pressure-Engineering Excitonic Properties of Two-Dimensional Hybrid Perovskites

Tingting Yin

Abstract

Two-dimension (2D) hybrid organic-inorganic perovskites (HOIPs) are formed naturally multiple-quantum-well structures with a much larger carrier binding energy, which possess stable excitons even at room temperature. In addition, 2D HOIPs allow us to exfoliate them into ultrathin flakes and stack them into various heterostructures, extending their photophysical properties. Therefore, 2D HOIPs are promising candidates for optoelectronic device applications, such as light-emitting diodes, lasing, etc. In this chapter, a summary of the crystal structures of 2D HOIP crystals and their heterostructures, excitonic properties, and the current research progress of the 2D HOIPs and their heterostructures are revealed. Next, high-pressure technology will be studied in detail on the effective engineering of crystal structures and exciton properties of 2D HOIPs toward significantly optimizing their functionalities. Finally, a summary is given, and the high-pressure strategy toward manipulation of 2D perovskite-based heterostructures is rationalized for next-generation high-performance excitonic devices.

Keywords: 2D HOIPs, perovskite heterostructures, exciton, high pressure, photophysical properties and crystal structures

1. Introduction

Two-dimensional (2D) hybrid organic-inorganic perovskite (HOIPs) have been star materials due to their excellent optoelectronic properties [1, 2], chemical tunability [3], and environmental stability [4, 5]. 2D HOIPs are composed of alternating layers of organic and inorganic parts, which undergo self-assembly in solution to generate regularly arranged stacked with organic and inorganic components. 2D HOIPs present a large compositional and structural phase space as well as exciting physical properties, such as a soft and dynamic structure [6], strong anisotropy in the in-plane and out-of-plane directions [7], strong light harvesting capability [8], high photoluminescence quantum yield (PLQY) [9], controllable charge-carrier mobility [10], stable excitons at room temperature [11–14], long carrier diffusion length [15, 16], and nonlinear effects [17–20], etc. Breakthrough performances in 2D HOIP-based devices have been achieved since 2016, including solar cells [21–23], light-emitting diodes (LEDs) [24–27], lasing [28–31] and photodetectors [32–37]. On the other hand, for integrating 2D HOIPs into electronic devices, precise control over layer thickness is crucial. The modest interaction among organic chains within unit cells facilitates the mechanical exfoliation of thin flakes of 2D HOIPs, employing techniques akin to those developed for graphene and other 2D layered materials [38, 39]. In these exfoliated perovskite flakes, alterations are expected in exciton-binding energy, dielectric constant, and spatial extension of electron and hole wavefunctions, particularly as thickness diminishes toward the atomic scale. Researchers have investigated the nonlinear optical (NLO) effects, such as third-harmonic generation (THG), in these exfoliated thin flakes with thickness in the 20–60 nm range [40, 41]. Furthermore, surface and interface effects become significantly more influential in these molecularly ultrathin flakes, exerting a profound impact on their photophysical and electronic characteristics [42, 43]. For example, surface disorder or lattice reconstructions in these molecularly thin perovskite flakes may disrupt surface-inversion symmetry, potentially resulting in Rashba-band splitting [38]. In addition, 2D HOIPs have become a strong candidate recently as a new building block for fabricating various heterostructures with extended physical properties. Until now, 2D perovskite-based vertical and lateral heterostructures have been fabricated based on multiple methods, such as dry transfer, vapor deposition, and solution-based chemical synthesis [44–47]. Through modulating materials and the stacking sequence integrated with 2D HOIPs and other 2D materials, various 2D perovskite-based heterostructures have been obtained, which not only provides a unique platform for researchers to explore new fundamental physics but also expands the potential applications of novel optoelectronic devices [47–51]. It is worth noticing that the novel physical properties observed in 2D HOIP crystals have close relationships with their microscopic crystal structures. For example, a tiny change in their crystal structures, such as bond length and bond angle, will induce a giant change in their optical and electronic properties. Therefore, comprehensively understanding their crystal structures and resolving the structure-property correlations at the atomic level is important for further improving their material properties and device functionalities. Moreover, for the development of heterostructure devices utilizing 2D HOIPs, a comprehensive grasp of exciton interactions, as well as charge and energy transfer mechanisms occurring at the interfaces between 2D layers, is imperative [48].

Pressure engineering as an effective and clean technique can systematically modify the crystal structures and electronic states of solid materials, resulting in new materials, new properties, and new physics that cannot be obtained by using other methods. Particularly, HOIPs with a variety of compositions have soft lattices, where the bond length and bond angle of the inorganic framework will be drastically changed under pressure, significantly modulating their crystal structures and photophysical properties. Therefore, *in situ* high-pressure investigations provide researchers with a unique platform to obtain a comprehensive understanding of the structure-property relationship of HOIPs. Previous extensive studies have demonstrated that excitonic emission behaviors of 2D HOIPs can be effectively manipulated by external hydrostatic pressure, including the emission energy, intensity, linewidth, and lifetime [52]. The physical origin for triggering these excitonic emission changes is the pressure-induced complex crystal structural phase transitions of 2D HOIPs.

In this review, we will discuss the current understanding of the crystal structures and physical properties of 2D HOIPs with a particular focus on various crystal structures, heterostructures, excitonic band structures, and light–matter interactions in these materials. Particularly, high-pressure technique will be mainly discussed as an effective tool for the manipulation of crystal structures and excitonic properties of 2D HOIPs. In the final part, we will give a summary, and also provide a perspective on the high-pressure engineering of 2D perovskite-based heterostructures, addressing key scientific questions in the physical properties that are required to be resolved to realize breakthroughs in their functionalities for next-generation high-performance optoelectronic device applications.

2. Crystal structure and physical properties of 2D HOIPs

2.1 Crystal structure of 2D HOIPs

The general chemical formula of 2D HOIPs is $(RNH_3)_2A_{n-1}B_nX_{3n+1}$, where n represents the number of inorganic layers. RNH_3^+ is a long-chain alkylammonium organic molecular layer, such as butyl ammonium (BA) and phenylethyl ammonium (PEA), A is a smaller cation filled into the 12-fold coordinated holes wrapped by the BX_6 octahedra, such as $CH_3NH_3^+$ (MA^+), formamidinium (FA), B refers to metal cations, usually Pb or Sn, and X stand for halide anion (**Figure 1a**) [55–57]. 2D HOIPs can be structurally considered as slicing three-dimensional (3D) perovskite frameworks by inserting organic cation spacers among the BX_6 inorganic framework along the (100), (110), and (111) different crystallographic orientations (**Figure 1b**) [4, 6, 53]. The conversion from pure 2D perovskites to quasi-2D perovskites and then to 3D perovskites will be realized by increasing the number (n) of inorganic layers (**Figure 1a**) [4, 6, 7]. The extensively studied candidate of 2D HOIPs is the (100) perovskite derivatives, such as Ruddlesden-Popper (RP) phase and Dion-Jacobson (DJ) phase (**Figure 1b**) [7, 53, 58–60]. For the RP phase perovskites, the adjacent organic layers are connected via weak Van der Waals (vdW) forces, while there is only one layer of the organic molecules in the DJ phase perovskites (**Figure 1c**) [54]. Due to the interaction between the organic and inorganic layers occurring via hydrogen bonding, DJ phase 2D perovskites demonstrate superior structural stability compared to RP phase 2D perovskites. Consequently, RP 2D perovskites are more readily mechanically exfoliated and stacked to form diverse 2D heterostructures.

2.2 Excitonic properties of 2D HOIPs

2D HOIPs naturally form structures akin to multiple-quantum-wells, with adjacent inorganic layers sandwiched by two layers of long-chain organic spacer cations. Specifically, the inorganic $(BX_6)^{4-}$ layers act as the "wells" while the organic spacers act as the "barriers", giving rise to these quantum-well configurations (**Figure 2a**) [61, 63]. Such a setup results in large dielectric confinement within 2D HOIPs due to the substantial difference in dielectric constants between the organic layer barriers and the inorganic layers. Consequently, there are huge oscillator strengths and large exciton-binding energy (> 100 meV) [7]. As a result, electrons and holes are strongly confined within the 2D inorganic slabs, leading to stable excitonic emission even at room temperature (**Figure 2b**) [61]. Given that quantum and dielectric confinement effects strongly rely on the number of inorganic layers (n), the band structure of 2D HOIPs can be efficiently modulated by adjusting this parameter. For example, increasing the value of n results in thicker unit cells, accompanied by a

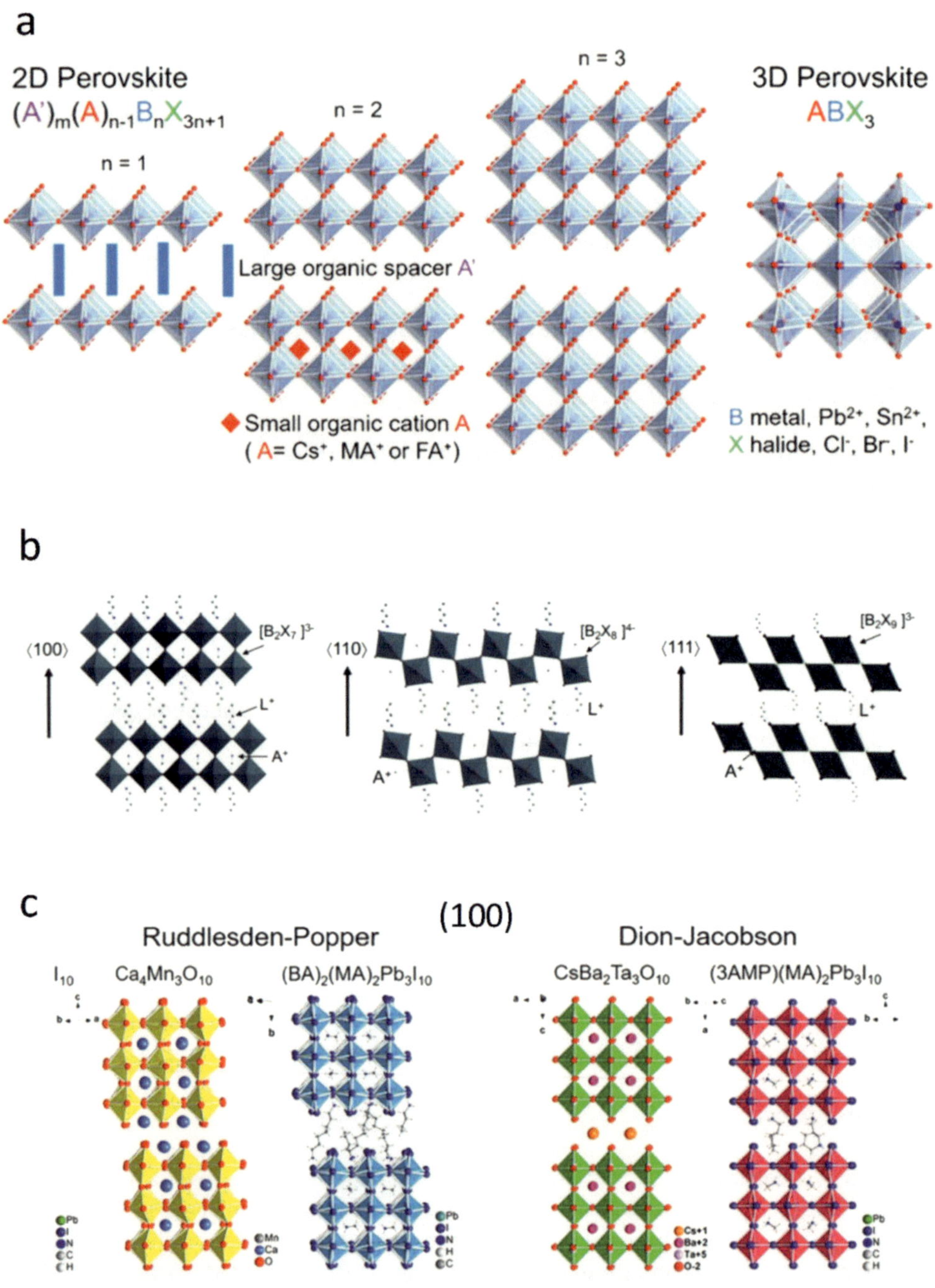

Figure 1.
(a) schematic illustration of the evolution from 2D perovskite to 3D perovskite with key components [6]. Copyright 2018 American Chemical Society, (b) schematic structures of different oriented families of 2D perovskites along (100), (110), and (111) planes, respectively [53]. Copyright 2019 MDPI, and (c) structural details of PR phase and DJ phase 2D perovskite, respectively [54]. Copyright 2019 Elsevier.

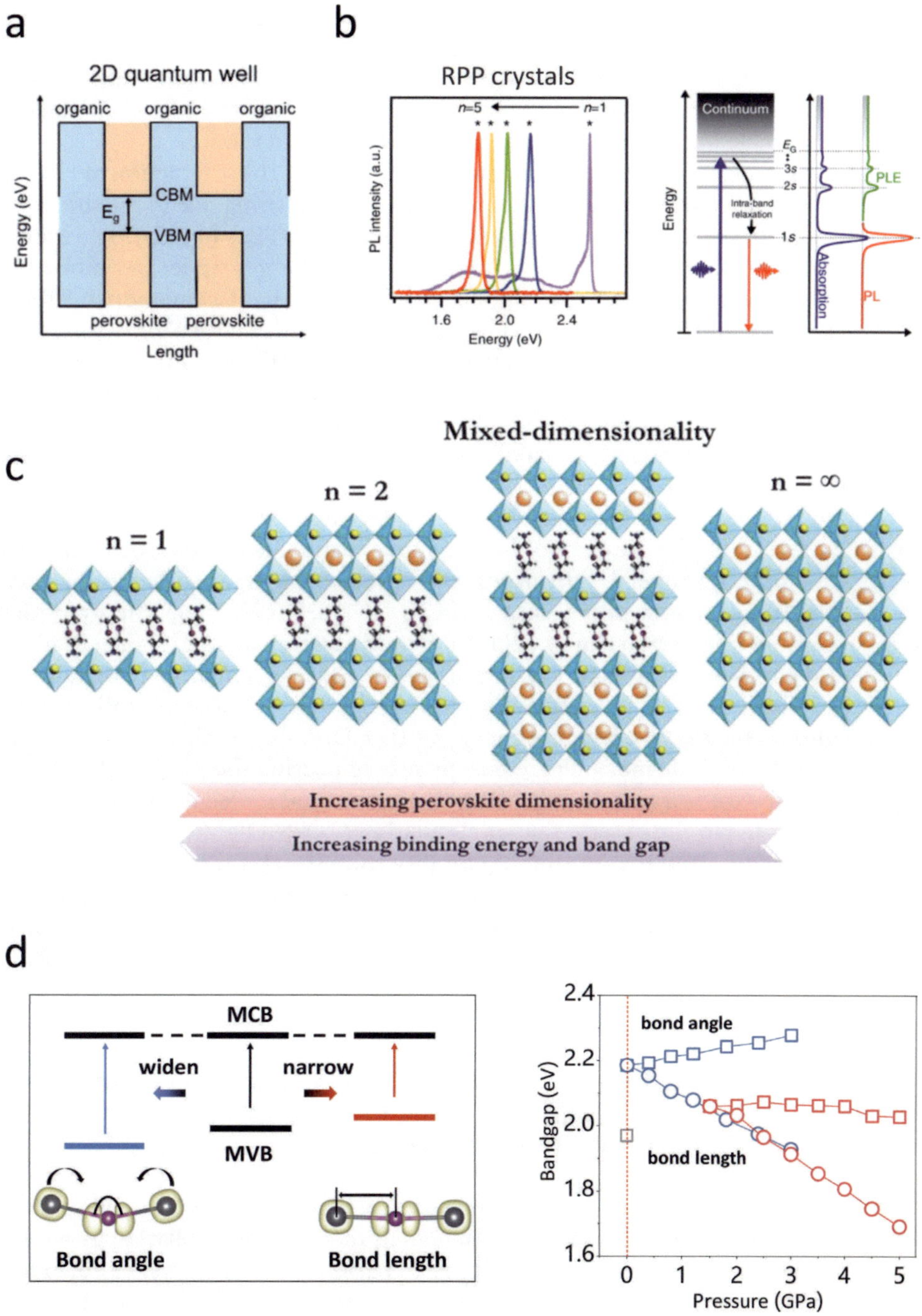

Figure 2.
(a) 2D quantum well structure illustration of 2D HOIPs [6]. Copyright 2018 American Chemical Society, (b) excitonic emission (left) and schematics of the Rydberg of the exciton ground states merging with the continuum (right) [61]. Copyright 2018 springer nature, (c) diagram showing the binding energy and band gap vs. layer number of n [2]. Copyright 2016 advanced materials, and (d) correlation of structure parameter of bond angle and bond length on the excitonic band structure of 2D HOIP [62]. Copyright 2018 American Chemical Society.

reduction in the band gap and excitonic binding energies (**Figure 2c**) [1]. Moreover, the electronic band structure of 2D HOIPs is intricately linked with the crystal structural distortion of inorganic sublattices, specifically the bond angle of B-X-B and bond length of B-X. These distortions induce shit in the valence band and conduction bands, thereby causing variations in the bandgap (**Figure 2d**) [62].

Furthermore, the electron-phonon interaction, namely, the interaction of charge carriers with lattice vibrations (phonons), in 2D HOIPs is strong due to the soft lattice nature, leading to the formation of self-trapped excitons (STEs) [64, 65]. The states of STEs are below the intrinsic band edge which affects charge-carrier recombination pathways and dynamics (**Figure 3a**) [67, 68, 70]. Particularly, the emission of STEs is the physical origin of the intensive broadband emission with a large Stokes shift and even the white-light emission observed in some 2D HOIPs (**Figure 3b**) [68]. In 2014, there were reports of three-layered perovskites producing broadband white-light emission when excited by near-UV light. These perovskites showcased an impressively high color rendering index, making them particularly suitable for applications in natural white LEDs [67].

Although the electronic band structure of 2D HOIPs is primarily governed by the inorganic sublattices, the presence of long organic ligands also holds significance in altering their optical properties. Incorporating diverse long organic chains opens avenues for exploring a wealth of physics within 2D HOIPs. For example, the strength of electron-phonon interaction can be adjusted by varying the organic ligands, leading to enhanced emission of STEs surpassing that of free exciton emission in certain 2D HOIPs (**Figure 3c**) [70, 71]. Secondly, altering the organic molecules allows for manipulation of the exciton-binding energy in 2D HOIPs due to differing dielectric constants among organic molecules, consequently impacting the PLQY [72, 73]. Thirdly, the substitution of long organic chains with chiral molecules in specific 2D HOIPs has resulted in the observation of circularly polarized light emission. This phenomenon arises from the transfer of chirality from the chiral molecule to the 2D perovskites (**Figure 3d**) [69, 74–76]. Such 2D HOIPs exhibit circular dichroism (CD) on account of the different absorption of left and right-handed circularly polarized light, holding significant promise for applications in spintronics and valleytronics [75, 77–79].

3. Fabrication and excitonic properties of 2D perovskite heterostructures

3.1 Fabrication of 2D perovskite heterostructures

The inherent layered structure and weak interlay interactions of 2D HOIPs, particularly in RP phase perovskites, facilitate their mechanical exfoliation using the conventional tape method, enabling the transformation from bulk crystals down to molecularly thin forms (**Figure 4a**) [38, 80]. Such thin and soft forms of 2D HOIPs can be easily picked up and transferred onto various substrates, including 2D HOIPs and other 2D layered materials, such as 2D transition metal dichalcogenides (TMDCs) and graphene, to fabricate vertically stacked artificial heterostructures. In addition, researchers have employed solution-based methods (bottom-up approaches) to naturally generate both in-plane and vertical 2D perovskite-based heterostructures.

As an illustration, researchers have successfully produced 2D RP phase perovskite multi-heterostructures comprising different layer numbers of n using a simple solution growth technique. This method has enabled the creation of vertically stacked

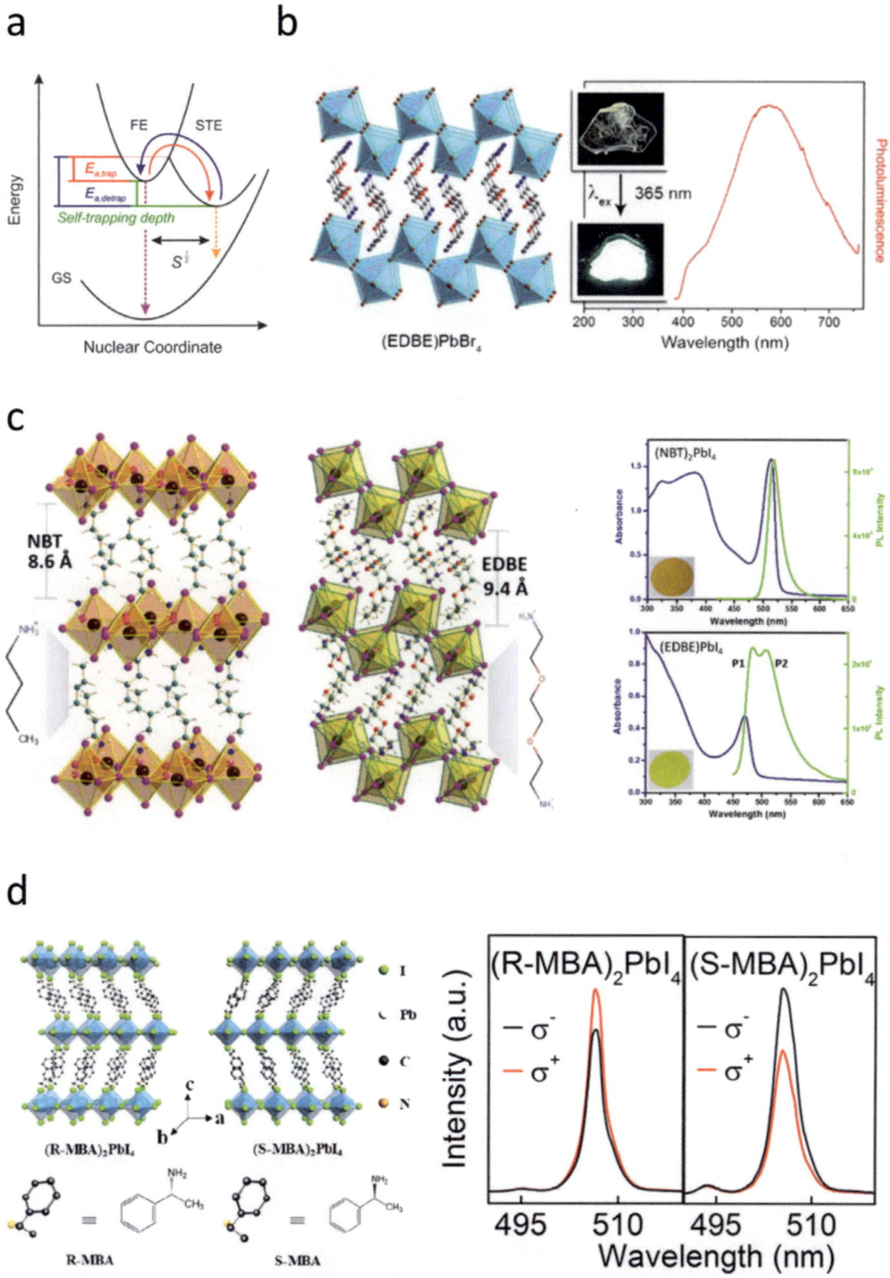

Figure 3.
(a) Nuclear coordinate diagram illustrating the dynamic process of exciton self-trapping and de-trapping a 2D perovskite, with labels indicating different states and activation energies. (GS = ground state, FE = free exciton state, STE = self-trapped exciton state, $E_{a,trap}$ = activation energy for self-trapping, $E_{a,detrap}$ = activation energy for de-trapping, S=Huang-Phys parameter) [66]. Copyright 2017 the Royal Society chemistry, (b) broadband emission covering the entire visible spectrum from (110) 2D perovskite, accompanied by photographic images of the crystal in the inset [67]. Copyright 2014 American Chemical Society, (c) 2D perovskites with short and long organic ligands (left), showcasing the corresponding narrow and broad PL emission (right). The insets demonstrate the structure-induced color change (from orange to yellow) [68]. Copyright 2016 American Chemical Society, and (d) schematic illustrations of the crystal structures of 2D perovskites with R- and S-MBA chiral organic molecular chains, along with circularly polarized PL spectra of these chiral 2D perovskites at 77 K [69]. Copyright 2019 American Chemical Society.

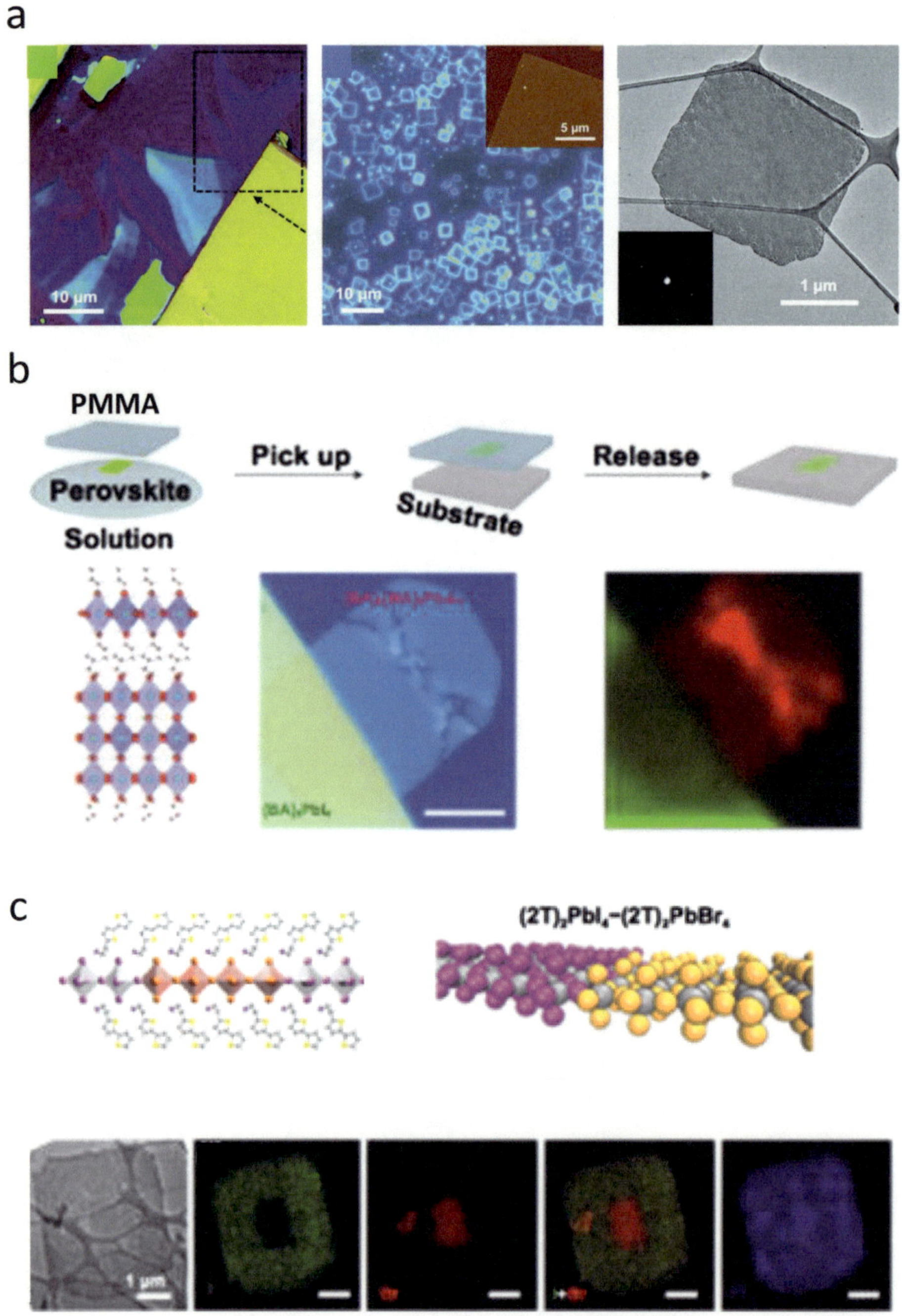

Figure 4.
(a) Mechanically exfoliated $((C_4H_9NH_3)_2PbI_4)$ *n = 1 2D RPP nanocrystals, the optical image of the exfoliated perovskite nanoflakes (left), the dark field optical image of the exfoliated perovskite nanoflakes (middle), and the AFM image of the monolayer perovskite sheet (right) [80]. Copyright 2017 the Royal Society of Chemistry, (b) construction of 2D perovskite heterostructures with dry transfer method (top) and the optical and PL images of the staked perovskite heterostructure(bottom) [48]. Copyright 2023 IOP publishing ltd, and (c) schematic illustrations of the* $(2\ T)_2PbI_4/(2\ T)_2PbBr_4$ *lateral 2D perovskite heterostructures (top) and the corresponding TEM image and EDS mappings of the 2D perovskite heterostructures (bottom) [48]. Copyright 2023 IOP publishing ltd.*

double heterostructures and intricate multilayer heterostructures of 2D lead iodide perovskites through vdW epitaxy [81]. Furthermore, they have achieved controllable direct growth of large-area nanosheets of various phase-pure RP perovskites with thickness down to monolayer. These free-standing perovskite nanosheets, assembled at the air-liquid interface, were lifted using a PDMS stamp and transferred onto Si/ SiO_2 substrates. Subsequently, a second layer of free-standing perovskite nanosheet was stacked on top to form a heterostructure. By iterative pick-up and transfer steps, they have systematically assembled these nanosheets into diverse 2D perovskite heterostructures (**Figure 4b**) [48, 82]. In addition, in-plane 2D perovskite heterostructures have also been obtained by a sequential solution phase approach, where two adjoined perovskites will be epitaxially grown in the lateral direction from the center to the edge. They have been demonstrated highly stable a tunable superlattice (**Figure 4c**) [44, 48, 83].

3.2 Optical properties of 2D perovskite heterostructures

2D perovskite heterostructures incorporating diverse layered configurations offer distinct optical properties, thereby significantly broadening the scope of potential functionalities and applications in devices. Firstly, the varying electronic band structures of 2D perovskite series with different layer numbers of n facilitate energy transfer from the higher energy bandgaps (lower n values) to the lower energy bandgaps (high n values), naturally giving rise to different types of band alignments [84]. For example, in the vertical heterostructure of $(PEA)_2PbI_4/(PEA)_2(MA)Pb_2I_7$ ($n = 1/n = 2$), energy transfer from $(PEA)_2PbI_4$ to $(PEA)_2(MA)Pb_2I_7$ occurs within hundreds of picoseconds after photoexcitation, resulting in the formation of a type I band alignment (**Figure 5a**) [81]. More importantly, charge-transfer excitons (CTEs) could be effectively formed in 2D perovskite heterostructures, as demonstrated in the PEA_2PbI_4:PEA_2SnI_4 heterostructure with the host and guest components, where broad PL occurs in the below-bandgap region (**Figure 5b**). These CTEs, crucial states formed at interfaces, introduce a novel mechanism for enhancing the multifunctionality of 2D perovskites [81, 85, 87]. Secondly, the interface containing edge-contact organic cations, chirality and symmetry properties allows for the property transfer across the interface by selecting different organic components, thus leading to the emergence of new physics in the interfaces. For example, the chirality can be transferred to the inorganic layers that are interfaced with the chiral molecules, providing a means to break valley or spin degeneracy within the inorganic layer [86]. Additionally, certain 2D perovskite heterojunctions exploit surface-bound ligands as spatial barriers to inhibit ion migration across the junctions, enabling multicolor emission with high spectral purity [81]. Also, the investigations into giant nonlinear optical responses have been conducted in $(n\text{-}C_4H_9NH_3)_2PbI_4/(n\text{-}C_4H_9NH_3)_2(CH_3NH_3)Pb_2I_7$ heterostructure of centimeter size, revealing giant two-photon absorption and strong multiphoton-induced PL in the heterostructure of varying thickness (**Figure 5c**) [88]. In the lateral epitaxial 2D perovskite heterostructures, band alignments can be modified either by varying the inorganic composition in the lateral in-plane direction or by altering the molecular structure in the out-of-plane direction, thereby determining PL emission properties. For example, a type-II band alignment is achieved in $(4Tm)_2SnI_4$–$(4Tm)_2PbI_4$ heterostructure, where exciton emission predominates at the edge and the lifetime is enhanced at the interface of the heterostructure [83].

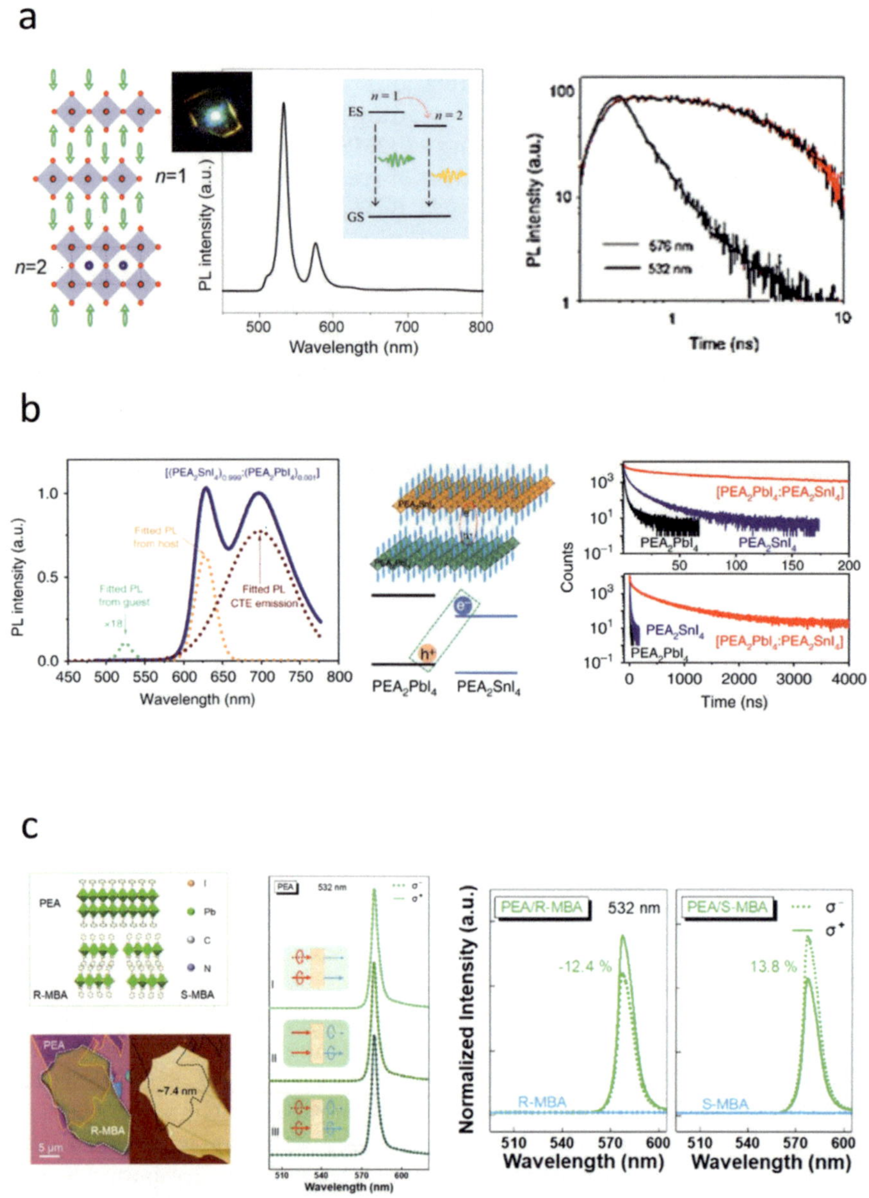

Figure 5.
(a) Schematic structure illustrating 2D heterostructure composed of RPP with n = 1 and n = 2 (left), along with the corresponding PL spectrum exhibiting two emission peaks at 532 nm and 576 nm assigned to n = 1 and n = 2 RPP (middle). The inset depicts the band alignment within the heterostructure, showcasing energy transfer from n = 1 to n = 2 layer. GS: Ground state, ES: Excited state. Lifetime curves of the heterostructure probed at both components' emission wavelengths are presented on the right side [81]. Copyright 2018 American Chemical Society, (b) PL spectrum of a 2D perovskite heterostructure, demonstrating a broad emission with the peak position at 697 nm, alongside intrinsic emission peaked at 524 and 627 nm from host PEA_2PbI_4 and guest PEA_2SnI_4 components (left). The middle schematic diagram illustrates the formation of CTEs at the interfaces between these 2D perovskite heterostructures. PL lifetimes of the broad emission at 669 nm, intrinsic emission at 524 nm, and 621 nm of this 2D perovskite heterostructures are depicted on the right side [85]. Copyright 2020 springer nature, and (c) schematic illustration of the crystal structures of PEA (non-chiral) and R-/S-MBA perovskites and optical/AFM images of this 2D heterostructure with the PEA layer on top of the R-MBA layer (left). Circularly polarized PL (CPL) spectra of pure PEA exhibit no chirality (middle), whereas CPL spectra of PEA/R-MBA and PEA/S-MBA heterostructures exhibit clear chirality (right) [86]. Copyright 2022 American Chemical Society.

4. High pressure

4.1 High-pressure technique

The diamond-anvil cell (DAC) stands out as a widely employed technique for generating ultra-high hydrostatic pressure on a small number of samples [89, 90]. A typical DAC utilized in laboratory settings comprises several essential components (**Figure 6a**). At the heart of a DAC lies a pair of diamonds situated on opposing piston and cylinder components, enabling the generation of high pressure at the pointed ends of each diamond by mutual compression facilitated by the metal piston and cylinder (**Figure 6b**). Sandwiched between the diamond pieces is a stainless-steel gasket featuring a central drilled hole, serving as the sample chamber (**Figure 6b**) [91]. To ensure effective and uniform transmission of pressure to the sample, the sample space is typically filled with a pressure-transmitting medium, commonly a liquid (**Figure 6b**). During pressurizing, the pressure within the liquid escalated significantly and was evenly transmitted to the sample in all directions, resulting in hydrostatic pressure. For perovskite samples, silicone oil is frequently employed as the pressure-transmitting medium [93]. The PL peak position of ruby serves as a reference for pressure calibration, exhibiting a nearly linear shift with hydrostatic pressure below 100 GPa (**Figure 6c**) [92, 94].

4.2 High-pressure engineering of excitonic behaviors in 2D HOIPs

Over the past 6 years, there has been considerable research into the impact of pressure on the structure and excitonic properties in 2D HOIPs. Controlled compression-induced modulation of the inorganic sublattices in these materials can finely adjust their photophysical properties. Specifically, lattice compression can effectively alter the excitonic states by narrowing or broadening the band gap and introducing new exciton bands. This presents a promising avenue for engineering 2D HOIPs with tailored properties under ambient pressure conditions. For example, the exciton emission peak is effectively tuned in the mechanically exfoliated thin flakes of $(BA)_2PbI_4$ (n = 1) 2D perovskite, which exhibits a blue shift at very low pressure of 0.1 GPa along with a continuous redshift up to 5.3 GPa before the occurrence of amorphization, the total energy change of excitonic emission is about 350 meV within a moderate pressure of 5.3 GPa (**Figure 7a**) [62]. The emission lifetimes of excitons have also changed accordingly, which is prolongated by 40 ps within 0.4 GPa and is shortened by 100 ps up to 2.3 GPa (**Figure 7a**) [62]. Besides, the emission linewidth of exciton has been investigated in the exfoliated ultrathin flakes of $(BA)_2(MA)Pb_2I_7$ (n = 2) 2D perovskite, where the narrow band emission is replaced by a stable broadband emission after pressure-treatment beyond certain pressure threshold (>3 GPa) due to the pressure-induced permanent distortion of inorganic sublattices and the formation of STEs (**Figure 7b**) [52]. Specifically, free excitons are firmly trapped by strongly distorted inorganic sublattices to form STEs, which cannot revert to free excitons due to the large lattice deformation potential induced by the pressurization process, leading to broadband emission at ambient conditions. Moreover, the emission intensity of excitons in $(HA)_2(GA)Pb_2I_7$ (HA = n-hexylammonium, GA = guanidinium) 2D perovskite has undergone a remarkably 12-fold enhancement. This enhancement is attributed to the substantial suppression of carrier trapping resulting from lattice compression under mild pressure conditions, consequently leading to

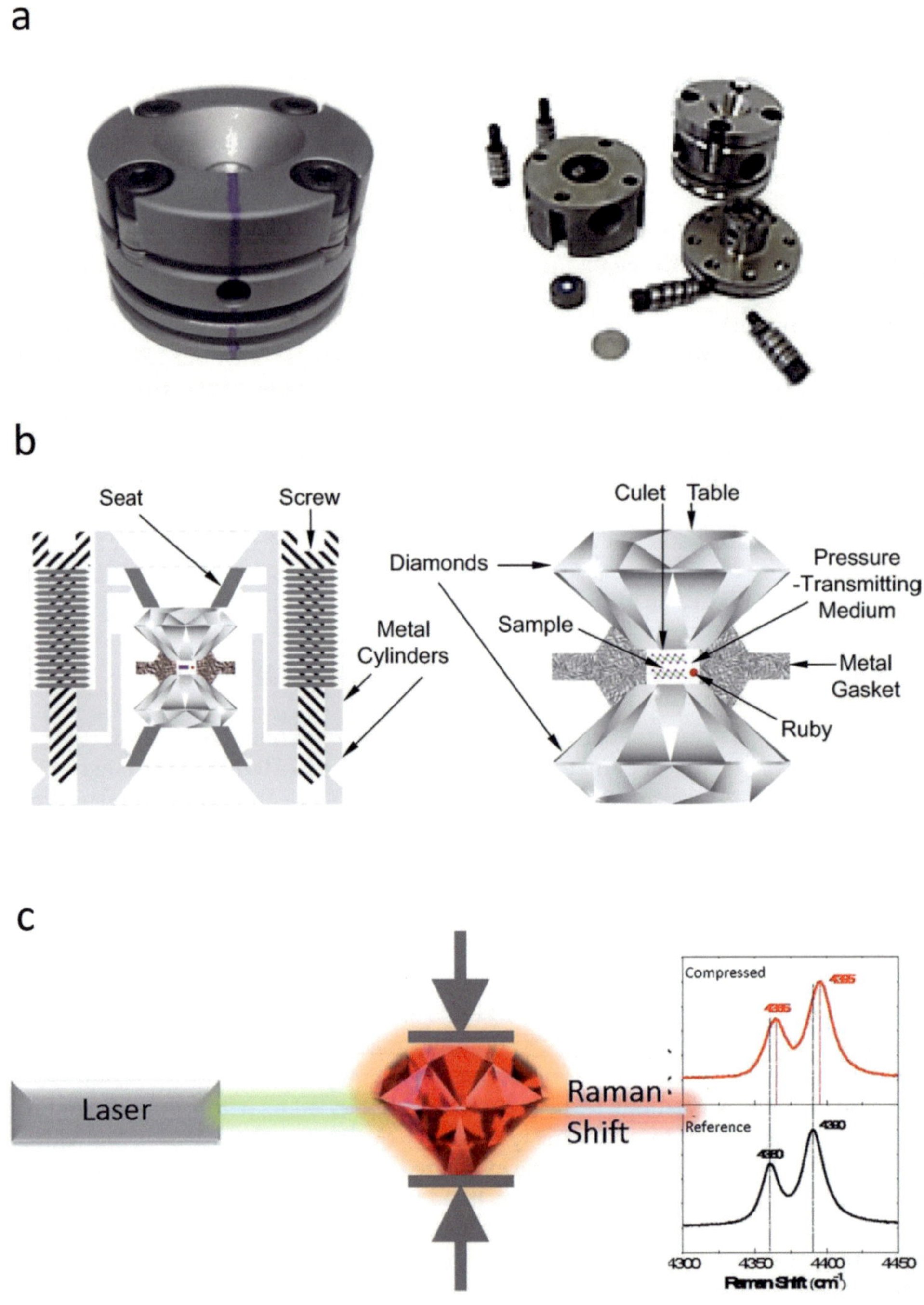

Figure 6.
(a) Actual images of a DAC, along with the metal piston, cylinder, and screws, are shown on the right side [91]. Copyright 2021 Elsevier ltd., (b) schematic illustration of the cross-section of a DAC (left), with a zoomed-in view (right) on the diamond/gasket assembly [91]. Copyright 2021 Elsevier ltd., and (c) typical fluorescence spectra of the ruby emission observed within the DAC under compression [92]. Copyright 2022 American Chemical Society.

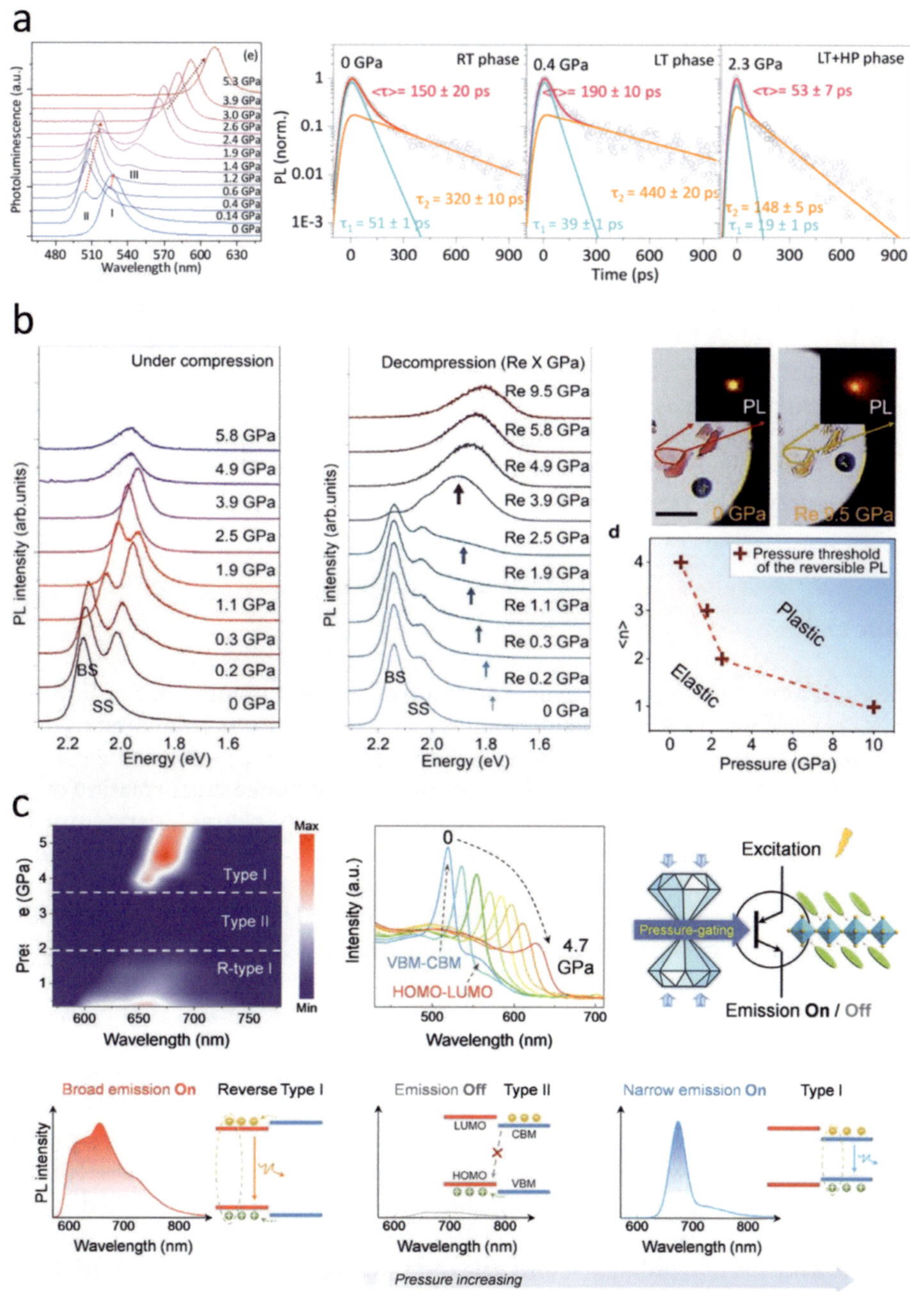

Figure 7.
(a) Evolution of the PL peak positions (left) and lifetimes (right) of 2D RPP (n = 1) under high pressure [62]. Copyright 2018 American Chemical Society, (b) In situ compression-decompression PL spectra of 2D RPP (n = 2), where broadband emission is observed upon releasing pressures from certain pressure points (left). The gradient-colored blue in the left figure arrows represent the proportion of the broadband emission. On the right side, optical photographs and fluorescence images of n = 2 RPP exfoliated flakes before and after pressure treatments are presented, with a scale bar of 50 μm. A plot summarizing the threshold pressure across from elastic to plastic regime for n = 1 to 4 RPPs (right and bottom) [52]. Copyright 2023 springer nature. **c,** *pressure-controlled emission properties and the corresponding band alignments at different states, along with a schematic illustration of pressure-gated multiple emission states where the emission activation or deactivation can be controlled [78]. Arrows indicate the changes in the PL spectra with increasing pressure. Copyright 2022 American Association for the Advancement of Science.*

significantly heightened free exciton emission [95, 96]. In addition, the manipulation of band-edge states and charge distribution in organic semiconductor-incorporated 2D halide perovskites has been realized recently. This manipulation reveals a switching of band alignment at the organic-inorganic interface, offering control over the emission properties of 2D perovskites (**Figure 7c**) [78]. More interestingly, the anisotropic structural nature of 2D HOIPs has proved that the compression process is highly anisotropic [97]. For example, the quasi-uniaxial compression primarily squeezes the organic layers in the layer-to-layer direction in $(PEA)_2PbI_4$ 2D perovskite crystal in a low-pressure region. This compression effectively modulates the quantum confinement effect by 250 meV, owing to the significant reduction in thickness of the organic barrier layers [98]. Pressure-induced emission from initially non-emissive 2D perovskites has also been investigated. For example, the STE emission is observed at 2.5 GPa with a broad emission band and large Stokes shift in initially nonfluorescent $(BA)_4AgBiBr_8$ 2D perovskite, with further significant enhancement upon compression up to 8.2 GPa. Such pressure-induced emission in these 2D perovskites is attributed to a structural phase transition, namely, $[AgBr_6]^{5-}$ and $[BiBr_6]^{3-}$ inter octahedra tilt [99].

5. Conclusion

2D HOIPs emerge through the incorporation of sizable organic cations derived from 3D perovskites, presenting enhanced structural stability as compared to their 3D counterparts. The large binding energy, stemming from the large dielectric constant difference between the inorganic and organic layers, facilitates the formation of stable excitons in 2D HOIPs even at room temperature. This characteristic furnishes a distinctive platform for the development of optoelectronic devices through the manipulation of excitons. Undoubtedly, the intrinsic excitons formed within band structures function as the principle excited states, pivotal for the development of high-performance light-emitting properties in such 2D HOIPs. Moreover, the thickness of 2D HOIPs down to the atomic limit to be stacked with various heterostructures largely expands their material diversity, triggering the emergence of new physics and extending optoelectronic heterostructure devices. In this chapter, we have reviewed recent advances in 2D HOIPs in terms of crystal structures, heterostructures, electronic band structures, and excitonic properties. To achieve high-performance exciton-based light-emitting devices, extensive exploration of the structure-property correlation in this type of perovskites has been conducted to comprehensively understand and precisely establish the relationship between crystal structures and excitonic properties. Among these methods, the utilization of high pressure has been shown to provide a means to regulate the unit cell and interatomic spacing of materials, bypassing the necessity for new growth methods or processing while enabling the examination of materials properties *in situ*. Under high pressure, the inherent structural anisotropy in 2D HOIPs results in discernible pressure responses along both the out-of-plane and in-plane directions, thereby facilitating substantial manipulation of crystal structures and excitonic states in 2D HOIPs during pressurization.

Fundamentally, the manipulation of intrinsic excitons involving the formation probability, binding energy, and radiative/nonradiative recombination governs the light-emitting properties. It is noteworthy to observe the effective behaviors of CTE/energy transfer in 2D perovskite heterostructures. This donor-acceptor design presents an intriguing aspect, where CTEs/energy transfers can be conveniently adjusted across energy, polarization, and spin parameters by artificially stacking different 2D perovskites together. This phenomenon gives rise to novel physical phenomena and unique capabilities to regulate optical, electrical, and magnetic functionalities through CTEs/energy transfers. Due to their high sensitivity to interlay spacing, CTEs/energy transfers exhibit more tunability, making them well-suited for pressure engineering, thus facilitating the acquisition of robust interlayer excitons. Moreover, high pressure serves as a powerful tool capable of altering the vdW interactions between layers, layer-to-layer sliding, twisted angles, as well as crystal structures in 2D perovskite heterostructures. This transformative capability significantly controls the dynamics of intralayer excitons, interlayer excitons, and free carriers, thus facilitating researchers in systematically and comprehensively unraveling the processes of interlayer exciton formation, relaxation, and transport in 2D perovskite heterostructures. The urgent need for high-pressure engineering of 2D perovskite heterostructures is underscored, both for fundamental research and practical applications. It provides researchers with new opportunities for vdW interaction engineering, effectively forming/controlling interlayer excitons, triggering new exciton physics, and developing novel heterostructure excitonic devices with desired properties.

Acknowledgements

T. Yin gratefully acknowledges strong support from the President Postdoctoral Fellowship of Nanyang Technological University.

Conflict of interest

The authors declare no conflict of interest.

Author details

Tingting Yin
Division of Physics and Applied Physics, School of Physical and Mathematical Sciences, Nanyang Technological University, Singapore, Singapore

*Address all correspondence to: ttyin@ntu.edu.sg

References

[1] Chen Y, Sun Y, Peng J, Tang J, Zheng K, Liang Z. 2D ruddlesden–popper perovskites for optoelectronics. Advanced Materials. 2018;**30**:1703487. DOI: 10.1002/adma.201703487

[2] Chen X, Zhou H, Wang H. 2D/3D Halide perovskites for optoelectronic devices. Frontiers in Chemistry. 2021;**9**: 1-17. DOI: 10.3389/fchem.2021.715157

[3] Leung TL, Ahmad I, Syed AA, Ng AMC, Popović J, Djurišić AB. Stability of 2D and quasi-2D perovskite materials and devices. Communications Materials. 2022;**3**:63. DOI: 10.1038/s43246-022-00285-9

[4] Quan LN, Yuan M, Comin R, Voznyy O, Beauregard EM, Hoogland S, et al. Ligand-stabilized reduced-dimensionality perovskites. Journal of the American Chemical Society. 2016;**138**:2649-2655. DOI: 10.1021/jacs.5b11740

[5] Kim E-B, Akhtar MS, Shin H-S, Ameen S, Nazeeruddin MK. A review on two-dimensional (2D) and 2D-3D multidimensional perovskite solar cells: Perovskites structures, stability, and photovoltaic performances. Journal of Photochemistry and Photobiology C Photochemistry Reviews. 2021;**48**:100405. DOI: 10.1016/j.jphotochemrev.2021.100405

[6] Mao L, Stoumpos CC, Kanatzidis MG. Two-dimensional hybrid halide perovskites: Principles and promises. Journal of the American Chemical Society. 2019;**141**:1171-1190. DOI: 10.1021/jacs.8b10851

[7] Stoumpos CC, Cao DH, Clark DJ, Young J, Rondinelli JM, Jang JI, et al. Ruddlesden–popper hybrid lead iodide perovskite 2D homologous semiconductors. Chemistry of Materials. 2016;**28**:2852-2867. DOI: 10.1021/acs.chemmater.6b00847

[8] Tan Z, Wu Y, Hong H, Yin J, Zhang J, Lin L, et al. Two-dimensional $(C_4H_9NH_3)_2PbBr_4$ perovskite crystals for high-performance photodetector. Journal of the American Chemical Society. 2016;**138**:16612-16615. DOI: 10.1021/jacs.6b11683

[9] Quan LN, Zhao Y, García de Arquer FP, Sabatini R, Walters G, Voznyy O, et al. Tailoring the energy landscape in quasi-2D halide perovskites enables efficient green-light emission. Nano Letters. 2017;**17**:3701-3709. DOI: 10.1021/acs.nanolett.7b00976

[10] Kober-Czerny M, Motti SG, Holzhey P, Wenger B, Lim J, Herz LM, et al. Excellent long-range charge-carrier mobility in 2D perovskites. Advanced Functional Materials. 2022;**32**:2203064. DOI: 10.1002/adfm.202203064

[11] Blancon J-C, Even J, Stoumpos CC, Kanatzidis MG, Mohite AD. Semiconductor physics of organic-inorganic 2D halide perovskites. Nature Nanotechnology. 2020;**15**:969-985. DOI: 10.1038/s41565-020-00811-1

[12] Guo Z, Wu X, Zhu T, Zhu X, Huang L. Electron-phonon scattering in atomically thin 2D perovskites. ACS Nano. 2016;**10**:9992-9998. DOI: 10.1021/acsnano.6b04265

[13] Zhang F, Lu H, Tong J, Berry JJ, Beard MC, Zhu K. Advances in two-dimensional organic-inorganic hybrid perovskites. Energy & Environmental Science. 2020;**13**:1154-1186. DOI: 10.1039/C9EE03757H

[14] Seitz M, Magdaleno AJ, Alcázar-Cano N, Meléndez M, Lubbers TJ, Walraven SW, et al. Exciton diffusion in two-dimensional metal-halide perovskites. Nature Communications. 2020;**11**:2035. DOI: 10.1038/s41467-020-15882-w

[15] Shrestha S, Li X, Tsai H, Hou C-H, Huang H-H, Ghosh D, et al. Long carrier diffusion length in two-dimensional lead halide perovskite single crystals. Chem. 2022;**8**:1107-1120. DOI: 10.1016/j.chempr.2022.01.008

[16] Yang L, Shi J, Wu Y, Jin X, Bie T, Hu C, et al. Long carrier diffusion length and efficient charge transport in thick quasi-two-dimensional perovskite solar cells enabled by modulating crystal orientation and phase distribution. ACS Applied Energy Materials. 2022;**5**:8930-8939. DOI: 10.1021/acsaem.2c01401

[17] Han X, Zheng Y, Chai S, Chen S, Xu J. 2D organic-inorganic hybrid perovskite materials for nonlinear optics. Nano. 2020;**9**:1787-1810. DOI: 10.1515/nanoph-2020-0038

[18] Wang H, Fang C, Luo H, Li D. Recent progress of the optoelectronic properties of 2D ruddlesden-popper perovskites. Journal of Semiconductors. 2019;**40**:041901. DOI: 10.1088/1674-4926/40/4/041901

[19] Chen Z, Zhang Q, Zhu M, Chen H, Wang X, Xiao S, et al. In-plane anisotropic nonlinear optical properties of two-dimensional organic-inorganic hybrid perovskite. Journal of Physical Chemistry Letters. 2021;**12**:7010-7018. DOI: 10.1021/acs.jpclett.1c01890

[20] He Y, Li X, Li J, Huang J, Zhu H, Feng Y, et al. Dispersion of the third-order optical nonlinearities in 2D $(PEA)_2PbI_4$ perovskite film. Optics Express. 2023;**31**:34292-34299. DOI: 10.1364/OE.502036

[21] Lee J-W, Dai Z, Han T-H, Choi C, Chang S-Y, Lee S-J, et al. 2D perovskite stabilized phase-pure formamidinium perovskite solar cells. Nature Communications. 2018;**9**:3021. DOI: 10.1038/s41467-018-05454-4

[22] Shao M, Bie T, Yang L, Gao Y, Jin X, He F, et al. Over 21% efficiency stable 2D perovskite solar cells. Advanced Materials. 2022;**34**:2107211. DOI: 10.1002/adma.202107211

[23] Di Carlo A, Agresti A, Brunetti F, Pescetelli S. Two-dimensional materials in perovskite solar cells. JPhys Energy. 2020;**2**:031003. DOI: 10.1088/2515-7655/ab9eab

[24] Zhang L, Sun C, He T, Jiang Y, Wei J, Huang Y, et al. High-performance quasi-2D perovskite light-emitting diodes: From materials to devices. Light: Science & Applications. 2021;**10**:61. DOI: 10.1038/s41377-021-00501-0

[25] Sun C, Jiang Y, Cui M, Qiao L, Wei J, Huang Y, et al. High-performance large-area quasi-2D perovskite light-emitting diodes. Nature Communications. 2021;**12**:2207. DOI: 10.1038/s41467-021-22529-x

[26] Shin YS, Park CB, Adhikari A, Yoon YJ, Cho HW, Son JG, et al. Manipulated interface for enhanced energy cascade in quasi-2D blue perovskite light-emitting diodes. ACS Energy Letters. 2022;**7**:3345-3352. DOI: 10.1021/acsenergylett.2c01785

[27] Gao Y, Liu Y, Zhang F, Bao X, Xu Z, Bai X, et al. High-performance perovskite light-emitting diodes enabled by passivating defect and constructing dual energy-transfer pathway through functional perovskite nanocrystals.

Advanced Materials. 2022;**34**:2207445. DOI: 10.1002/adma.202207445

[28] Qin C, Sandanayaka ASD, Zhao C, Matsushima T, Zhang D, Fujihara T, et al. Stable room-temperature continuous-wave lasing in quasi-2D perovskite films. Nature. 2020;**585**:53-57. DOI: 10.1038/s41586-020-2621-1

[29] Fruhling C, Wang K, Chowdhury S, Xu X, Simon J, Kildishev A, et al. Coherent random lasing in subwavelength quasi-2D perovskites. Laser & Photonics Reviews. 2023;**17**:2200314. DOI: 10.1002/lpor.202200314

[30] Gao X, Lin J, Guo X, He G, Zou D, Ishii T, et al. Room-temperature continuous-wave microcavity lasers from solution-processed smooth quasi-2D perovskite films with low thresholds. Journal of Physical Chemistry Letters. 2023;**14**:2493-2500. DOI: 10.1021/acs.jpclett.3c00279

[31] Chen K, Zhang Q, Liang Y, Song J, Li C, Chen S, et al. Quasi-two dimensional ruddlesden-popper halide perovskites for laser applications. Frontiers of Physics. 2023;**19**:23502. DOI: 10.1007/s11467-023-1347-6

[32] Li Z, Hong E, Zhang X, Deng M, Fang X. Perovskite-type 2D materials for high-performance photodetectors. Journal of Physical Chemistry Letters. 2022;**13**:1215-1225. DOI: 10.1021/acs.jpclett.1c04225

[33] Qiao B-S, Wang S-Y, Zhang Z-H, Lian Z-D, Zheng Z-Y, Wei Z-P, et al. Photosensitive dielectric 2D perovskite based photodetector for dual wavelength demultiplexing. Advanced Materials. 2023;**35**:2300632. DOI: 10.1002/adma.202300632

[34] Li B, Huang X, Wu X, Zuo Q, Cao Y, Zhu Q, et al. High-performing quasi-2D perovskite photodetectors with efficient charge transport network built from vertically orientated and evenly distributed 3D-like phases. Advanced Functional Materials. 2023;**33**:2300216. DOI: 10.1002/adfm.202300216

[35] Li W, Wu Q, Lu L, Tian Y, Luo H, Yun Y, et al. Molecular engineering for sensitive, fast and stable quasi-two-dimensional perovskite photodetectors. Journal of Materials Chemistry C. 2023;**11**:3314-3324. DOI: 10.1039/D2TC04932E

[36] Choi E, Zhang Y, Soufiani AM, Lee M, Webster RF, Pollard ME, et al. Exploration of sub-bandgap states in 2D halide perovskite single-crystal photodetector. npj 2D Materials and Applications. 2022;**6**:43. DOI: 10.1038/s41699-022-00317-5

[37] Wang H, Sun Y, Chen J, Wang F, Han R, Zhang C, et al. A review of perovskite-based photodetectors and their applications. Nanomaterials. 2022;**12**:4390. DOI: 10.3390/nano12244390

[38] Leng K, Fu W, Liu Y, Chhowalla M, Loh KP. From bulk to molecularly thin hybrid perovskites. Nature Reviews Materials. 2020;**5**:482-500. DOI: 10.1038/s41578-020-0185-1

[39] Soe CMM, Nagabhushana GP, Shivaramaiah R, Tsai H, Nie W, Blancon J-C, et al. Structural and thermodynamic limits of layer thickness in 2D halide perovskites. Proceedings of the National Academy of Sciences of the United States of America. 2019;**116**:58-66. DOI: 10.1073/pnas.1811006115

[40] Abdelwahab I, Grinblat G, Leng K, Li Y, Chi X, Rusydi A, et al. Highly enhanced third-harmonic generation in 2D perovskites at excitonic

resonances. ACS Nano. 2018;**12**:644-650. DOI: 10.1021/acsnano.7b07698

[41] Abdelwahab I, Dichtl P, Grinblat G, Leng K, Chi X, Park I-H, et al. Giant and tunable optical nonlinearity in single-crystalline 2D perovskites due to excitonic and plasma effects. Advanced Materials. 2019;**31**:1902685. DOI: 10.1002/adma.201902685

[42] Leng K, Abdelwahab I, Verzhbitskiy I, Telychko M, Chu L, Fu W, et al. Molecularly thin two-dimensional hybrid perovskites with tunable optoelectronic properties due to reversible surface relaxation. Nature Materials. 2018;**17**:908-914. DOI: 10.1038/s41563-018-0164-8

[43] Shao Y, Gao W, Yan H, Li R, Abdelwahab I, Chi X, et al. Unlocking surface octahedral tilt in two-dimensional ruddlesden-popper perovskites. Nature Communications. 2022;**13**:138. DOI: 10.1038/s41467-021-27747-x

[44] Shi E, Yuan B, Shiring SB, Gao Y, Akriti GY, Su C, et al. Two-dimensional halide perovskite lateral epitaxial heterostructures. Nature. 2020;**580**:614-620. DOI: 10.1038/s41586-020-2219-7

[45] Park JY, Lee YH, Kim H, Dou L. Two-dimensional perovskite heterostructures for single crystal semiconductor devices. Journal of Applied Physics. 2023;**134**:060901. DOI: 10.1063/5.0153306

[46] Miao S, Liu T, Du Y, Zhou X, Gao J, Xie Y, et al. 2D material and perovskite heterostructure for optoelectronic applications. Nanomaterials. 2022;**12**:2100

[47] Karpińska M, Jasiński J, Kempt R, Ziegler JD, Sansom H, Taniguchi T, et al. Interlayer excitons in $MoSe_2$/2D perovskite hybrid heterostructures–the interplay between charge and energy transfer. Nanoscale. 2022;**14**:8085-8095. DOI: 10.1039/D2NR00877G

[48] Wang H, Chen Y, Li D. Two/quasi-two-dimensional perovskite-based heterostructures: Construction, properties and applications. International Journal of Extreme Manufacturing. 2023;**5**:012004. DOI: 10.1088/2631-7990/acab40

[49] Wei Q, Wen X, Hu J, Chen Y, Liu Z, Lin T, et al. Site-controlled interlayer coupling in WSe_2/2D perovskite heterostructure. Science China Materials. 2022;**65**:1337-1344. DOI: 10.1007/s40843-021-1911-6

[50] Miao S, Liu T, Du Y, Zhou X, Gao J, Xie Y, et al. 2D material and perovskite heterostructure for optoelectronic applications. Nanomaterials (Basel). 2022:**12**:2100. DOI: 10.3390/nano12122100

[51] Li C, Loh KP, Leng K. Organic-inorganic hybrid perovskites and their heterostructures. Matter. 2022;**5**:4153-4169. DOI: 10.1016/j.matt.2022.11.002

[52] Yin T, Yan H, Abdelwahab I, Lekina Y, Lü X, Yang W, et al. Pressure driven rotational isomerism in 2D hybrid perovskites. Nature Communications. 2023;**14**:411. DOI: 10.1038/s41467-023-36032-y

[53] Gai C, Wang J, Wang Y, Li J. The low-dimensional three-dimensional tin halide perovskite: Film characterization and device performance. Energies. 2020;**13**:2. DOI: 10.3390/en13010002

[54] Lekina Y, Shen ZX. Excitonic states and structural stability in two-dimensional hybrid organic-inorganic perovskites. Journal of Science: Advanced Materials and Devices. 2019;**4**:189-200. DOI: 10.1016/j.jsamd.2019.03.005

[55] Tsai H, Asadpour R, Blancon J-C, Stoumpos CC, Even J, Ajayan PM, et al. Design principles for electronic charge transport in solution-processed vertically stacked 2D perovskite quantum wells. Nature Communications. 2018;**9**:2130. DOI: 10.1038/s41467-018-04430-2

[56] Grancini G, Nazeeruddin MK. Dimensional tailoring of hybrid perovskites for photovoltaics. Nature Reviews Materials. 2019;**4**:4-22. DOI: 10.1038/s41578-018-0065-0

[57] Dou L, Wong AB, Yu Y, Lai M, Kornienko N, Eaton SW, et al. Atomically thin two-dimensional organic-inorganic hybrid perovskites. Science. 2015;**349**:1518-1521. DOI: 10.1126/science.aac7660

[58] Rahil M, Ansari RM, Prakash C, Islam SS, Dixit A, Ahmad S. Ruddlesden–popper 2D perovskites of type $(C_6H_9C_2H_4NH_3)_2(CH_3NH_3)_{n-1}Pb_nI_{3n+1}$ (n = 1-4) for optoelectronic applications. Scientific Reports. 2022;**12**:2176. DOI: 10.1038/s41598-022-06108-8

[59] Mao L, Ke W, Pedesseau L, Wu Y, Katan C, Even J, et al. Hybrid dion–jacobson 2D lead iodide perovskites. Journal of the American Chemical Society. 2018;**140**:3775-3783. DOI: 10.1021/jacs.8b00542

[60] Li X, Ke W, Traoré B, Guo P, Hadar I, Kepenekian M, et al. Two-dimensional dion–jacobson hybrid lead iodide perovskites with aromatic diammonium cations. Journal of the American Chemical Society. 2019;**141**:12880-12890. DOI: 10.1021/jacs.9b06398

[61] Blancon JC, Stier AV, Tsai H, Nie W, Stoumpos CC, Traoré B, et al. Scaling law for excitons in 2D perovskite quantum wells. Nature Communications. 2018;**9**:2254. DOI: 10.1038/s41467-018-04659-x

[62] Yin T, Liu B, Yan J, Fang Y, Chen M, Chong WK, et al. Pressure-engineered structural and optical properties of two-dimensional $(C_4H_9NH_3)_2PbI_4$ perovskite exfoliated nm-thin flakes. Journal of the American Chemical Society. 2019;**141**:1235-1241. DOI: 10.1021/jacs.8b07765

[63] Wang N, Cheng L, Ge R, Zhang S, Miao Y, Zou W, et al. Perovskite light-emitting diodes based on solution-processed self-organized multiple quantum wells. Nature Photonics. 2016;**10**:699-704. DOI: 10.1038/nphoton.2016.185

[64] Li J, Wang H, Li D. Self-trapped excitons in two-dimensional perovskites. Frontiers of Optoelectronics. 2020;**13**:225-234. DOI: 10.1007/s12200-020-1051-x

[65] Li H, Zhang X, Wang H, Yu J, Li K, Wei Z, et al. Optical characteristics of self-trapped excitons in 2D $(iso\text{-}BA)_2PbI_4$ perovskite crystals. Photonics Research. 2022;**10**:594-600. DOI: 10.1364/PRJ.444457

[66] Smith MD, Jaffe A, Dohner ER, Lindenberg AM, Karunadasa HI. Structural origins of broadband emission from layered Pb–Br hybrid perovskites. Chemical Science. 2017;**8**:4497-4504. DOI: 10.1039/C7SC01590A

[67] Smith MD, Karunadasa HI. White-light emission from layered halide perovskites. Accounts of Materials Research. 2018;**51**:619-627. DOI: 10.1021/acs.accounts.7b00433

[68] Cortecchia D, Neutzner S, Srimath Kandada AR, Mosconi E, Meggiolaro D, De Angelis F, et al. Broadband emission in two-dimensional hybrid perovskites: The role of structural deformation. Journal of the American Chemical Society. 2017;**139**:39-42. DOI: 10.1021/jacs.6b10390

[69] Ma J, Fang C, Chen C, Jin L, Wang J, Wang S, et al. Chiral 2D perovskites with a high degree of circularly polarized photoluminescence. ACS Nano. 2019;**13**:3659-3665. DOI: 10.1021/acsnano.9b00302

[70] Hu T, Smith MD, Dohner ER, Sher M-J, Wu X, Trinh MT, et al. Mechanism for broadband white-light emission from two-dimensional (110) hybrid perovskites. Journal of Physical Chemistry Letters. 2016;**7**:2258-2263. DOI: 10.1021/acs.jpclett.6b00793

[71] Dohner ER, Hoke ET, Karunadasa HI. Self-assembly of broadband white-light emitters. Journal of the American Chemical Society. 2014;**136**:1718-1721. DOI: 10.1021/ja411045r

[72] Gao Y, Shi E, Deng S, Shiring SB, Snaider JM, Liang C, et al. Molecular engineering of organic-inorganic hybrid perovskites quantum wells. Nature Chemistry. 2019;**11**:1151-1157. DOI: 10.1038/s41557-019-0354-2

[73] Filip MR, Qiu DY, Del Ben M, Neaton JB. Screening of Excitons by organic cations in quasi-two-dimensional organic-inorganic lead-halide perovskites. Nano Letters. 2022;**22**:4870-4878. DOI: 10.1021/acs.nanolett.2c01306

[74] Long G, Sabatini R, Saidaminov MI, Lakhwani G, Rasmita A, Liu X, et al. Chiral-perovskite optoelectronics. Nature Reviews Materials. 2020;**5**:423-439. DOI: 10.1038/s41578-020-0181-5

[75] Liu S, Kepenekian M, Bodnar S, Feldmann S, Heindl MW, Fehn N, et al. Bright circularly polarized photoluminescence in chiral layered hybrid lead-halide perovskites. Science Advances. 2023;**9**:eadh5083. DOI: 10.1126/sciadv.adh5083

[76] Li M, Fang F, Huang X, Liu G, Lai Z, Chen Z, et al. Chiral ligand-induced structural transformation of low-dimensional hybrid perovskite for circularly polarized photodetection. Chemistry of Materials. 2022;**34**:2955-2962. DOI: 10.1021/acs.chemmater.1c03622

[77] Ma J, Wang H, Li D. Recent progress of chiral perovskites: Materials, synthesis, and properties. Advanced Materials. 2021;**33**:2008785. DOI: 10.1002/adma.202008785

[78] Guo S, Li Y, Mao Y, Tao W, Bu K, Fu T, et al. Reconfiguring band-edge states and charge distribution of organic semiconductor–incorporated 2D perovskites via pressure gating. Science Advances. 2022;**8**:eadd1984. DOI: 10.1126/sciadv.add1984

[79] Ahn J, Ma S, Kim J-Y, Kyhm J, Yang W, Lim JA, et al. Chiral 2D organic inorganic hybrid perovskite with circular dichroism tunable over wide wavelength range. Journal of the American Chemical Society. 2020;**142**:4206-4212. DOI: 10.1021/jacs.9b11453

[80] Dou L. Emerging two-dimensional halide perovskite nanomaterials. Journal of Materials Chemistry C. 2017;**5**:11165-11173. DOI: 10.1039/C7TC02863F

[81] Fu Y, Zheng W, Wang X, Hautzinger MP, Pan D, Dang L, et al. Multicolor heterostructures of two-dimensional layered halide perovskites that show interlayer energy transfer. Journal of the American Chemical Society. 2018;**140**:15675-15683. DOI: 10.1021/jacs.8b07843

[82] Pan D, Fu Y, Spitha N, Zhao Y, Roy CR, Morrow DJ, et al. Deterministic fabrication of arbitrary vertical heterostructures of two-dimensional Ruddlesden–popper halide perovskites.

Nature Nanotechnology. 2021;**16**:159-165. DOI: 10.1038/s41565-020-00802-2

[83] Shi E, Dou L. Halide perovskite epitaxial heterostructures. Accounts of Materials Research. 2020;**1**:213-224. DOI: 10.1021/accountsmr.0c00047

[84] Zhang L, Zhang X, Lu G. Band alignment in two-dimensional halide perovskite heterostructures: Type I or type II? Journal of Physical Chemistry Letters. 2020;**11**:2910-2916. DOI: 10.1021/acs.jpclett.0c00376

[85] Zhang J, Zhu X, Wang M, Hu B. Establishing charge-transfer excitons in 2D perovskite heterostructures. Nature Communications. 2020;**11**:2618. DOI: 10.1038/s41467-020-16415-1

[86] Zhan G, Zhang J, Zhang L, Ou Z, Yang H, Qian Y, et al. Stimulating and manipulating robust circularly polarized photoluminescence in achiral hybrid perovskites. Nano Letters. 2022;**22**:3961-3968. DOI: 10.1021/acs.nanolett.2c00482

[87] Song Y, Zhang C, Liu W, Li X, Long H, Wang K, et al. High-efficiency energy transfer in perovskite heterostructures. Optics Express. 2018;**26**:18448-18456. DOI: 10.1364/OE.26.018448

[88] Wang J, Mi Y, Gao X, Li J, Li J, Lan S, et al. Giant nonlinear optical response in 2D perovskite heterostructures. Advanced Optical Materials. 2019;**7**:1900398. DOI: https: 10.1002/adom.201900398

[89] Qiao W, Sun H, Fan X, Jin M, Liu H, Tang T, et al. Interlayer coupling and pressure engineering in bilayer MoS_2. Crystals. 2022;**12**:693. DOI: 10.3390/cryst12050693

[90] Jayaraman A. Diamond anvil cell and high-pressure physical investigations. Reviews of Modern Physics. 1983;**55**:65-108. DOI: 10.1103/RevModPhys.55.65

[91] Pei S, Wang Z, Xia J. High pressure studies of 2D materials and heterostructures: A review. Materials and Design. 2022;**213**:110363. DOI: 10.1016/j.matdes.2021.110363

[92] Yang W, Cortés-Vega FD, Ahmadi K, Castaneda N, Paidpilli M, Majkic G, et al. Accurate ruby sensor for stress analysis in electronics. ACS Applied Electronic Materials. 2022;**4**:4332-4339. DOI: 10.1021/acsaelm.2c00588

[93] Ragan DD, Clarke DR, Schiferl D. Silicone fluid as a high-pressure medium in diamond anvil cells. The Review of Scientific Instruments. 1996;**67**:494-496. DOI: 10.1063/1.1146627

[94] Forman RA, Piermarini GJ, Barnett JD, Block S. Pressure measurement made by the utilization of ruby sharp-line luminescence. Science. 1972;**176**:284-285. DOI: 10.1126/science.176.4032.284

[95] Guo S, Zhao Y, Bu K, Fu Y, Luo H, Chen M, et al. Pressure-suppressed carrier trapping leads to enhanced emission in two-dimensional perovskite $(HA)_2(GA)Pb_2I_7$. Angewandte Chemie (International Ed. in English). 2020;**59**:17533-17539. DOI: 10.1002/anie.202001635

[96] Kong L, Liu G, Gong J, Mao L, Chen M, Hu Q, et al. Highly tunable properties in pressure-treated two-dimensional dion–jacobson perovskites. Proceedings of the National Academy of Sciences of the United States of America. 2020;**117**:16121-16126. DOI: 10.1073/pnas.2003561117

[97] Li H, Qin Y, Shan B, Shen Y, Ersan F, Soignard E, et al. Unusual pressure-driven phase transformation and band renormalization in 2D vdW hybrid lead

halide perovskites. Advanced Materials. 2020;**32**:1907364. DOI: 10.1002/adma.201907364

[98] Liu S, Sun S, Gan CK, del Águila AG, Fang Y, Xing J, et al. Manipulating efficient light emission in two-dimensional perovskite crystals by pressure-induced anisotropic deformation. Science Advances. 2019;**5**:eaav9445. DOI: 10.1126/sciadv.aav9445

[99] Fang Y, Zhang L, Wu L, Yan J, Lin Y, Wang K, et al. Pressure-induced emission (PIE) and phase transition of a two-dimensional halide double perovskite $(BA)_4AgBiBr_8$ ($BA=CH_3(CH_2)_3NH_3^+$). Angewandte Chemie. 2019;**58**:15249-15253. DOI: 10.1002/anie.201906311

Chapter 3

Impact of Panel Materials on Solar Cell Performance

Sivasami Kolandasamy and Thangalakshmi Sivalingam

Abstract

Modern household appliances rely on electricity for cleanliness, availability, and ease of use. Conserving energy is critical for economic and resource sustainability. Currently, the bulk of electricity is generated by coal-fired power plants. In addition to providing energy, they harm the environment by spewing flue gases, most notably carbon dioxide. The carbon footprint is the quantity of carbon dioxide emitted into the atmosphere as a result of the actions of a single individual, enterprise, or the general public, and there is a way to reduce it. Renewable and sustainable energy sources are crucial for producing power, in addition to energy conservation. Solar energy is a cost-effective, clean, and widely available renewable energy source. Photovoltaics (PV) is a simple and effective way to harness solar energy. PV cells, often known as solar cells, rely on solar radiation to produce energy. All solar radiation landing on a PV cell or panel is not totally converted into electrical energy; this is determined by the materials used in PV cells. Other elements that influence the performance of a solar panel include temperature, solar irradiation, and fill factor. This study aims to analyse and compare PV panel materials based on their characteristics and other relevant criteria.

Keywords: photovoltaic technology, solar energy, materials for solar panels, fill factor, solar panel performance

1. Introduction

Home energy use accounted for 26.9% of global power demand in 2018 as per International Energy Agency (IEA) 2020 [1], with consumption behaviours mostly influenced by household distinctive features. Household power usage has risen throughout the years as the human population has grown and living conditions have improved. The electrical power demand in metropolitan regions throughout the world is somewhat satisfied, but the goal remains unmet in many rural places. Electricity consumption has skyrocketed as a result of the rise in the artificial intelligence sector, which is now on level with traditional home and industrial consumers. According to the IEA, global energy consumption is predicted to increase at a quicker rate during the next 3 years, rising by an average of 3.4% per year until 2026. The organisation also anticipates that electricity demand from data centres, artificial intelligence (AI), and cryptocurrency-related businesses will double by 2026 [2].

Based on figures provided by the IEA, the electric power industry contributes significantly to planetary carbon emissions, constituting nearly 40% of total emissions in 2022. There is a longstanding connection between power use and carbon footprints, and this sector is the one that produces the most carbon emissions [3]. Carbon dioxide emissions grew dramatically from 1.7 to 33.1 Gt in 2018, owing to growing energy consumption [4]. The quantity of greenhouse gas emissions that a person, business, city, or state releases into the atmosphere is measured by their "carbon footprint." The footprint comprises both direct emissions from the combustion of fossil fuels and indirect emissions, such as those associated with the use of energy derived from fossil fuels. And, while it is often calculated in tonnes of carbon dioxide (CO2) per year, it really encompasses all gases that promote global warming. They consist of methane, nitrogen oxide, and chlorofluorocarbons. As environmental sustainability analyses shifted their attention to global warming and CO2 emissions, the carbon footprint emerged as a crucial instrument for tracking adherence to international emission reduction accords [5].

Carbon footprint calculation is a sophisticated assessment that takes many aspects into consideration and may be done on a variety of scales. In general, rather than being measured directly, the carbon footprint of a state, a firm, or a person is assessed using scientific criteria and international standards. The recommendations and methods for national greenhouse gas inventories by the Inter-Governmental Panel on Climate Change (IPCC) serve as the primary reference point. These guidelines give the typical criterion for estimating greenhouse gas emissions [6]. Minimising non-essential energy usage, mainly home electricity, is often regarded as one of the most promising and affordable methods to combat universal climate change and promote green growth, especially in industrialised nations.

Furthermore, using renewable energy sources will help to minimise carbon footprints. To completely assess the effects of renewable energy consumption on sustainability concerns, including individual carbon emissions and per capita ecological footprints, a study utilising data from 130 different countries was carried out between 1992 and 2019. Panel threshold regression models were created for this purpose. The results show that increasing renewable energy use helps to reduce carbon emissions. When renewable energy usage surpasses the threshold value, a 1% increase reduces per capita carbon emissions by 0.280–0.396% on a worldwide scale. Using renewable energy is one of the most effective methods to immediately reduce greenhouse gas emissions. Fossil fuel consumption may be reduced, lowering levels of greenhouse gases and slowing climate change. Alternative sources of energy, such as solar and wind power, can be extensively utilised. Renewable energy can also help to reduce dependency on restricted resources while promoting long-term economic prosperity [7]. Analogous research endorsed the utilisation of sustainable energy sources to mitigate carbon emissions [8–10].

Smart Power India (SPI) conducted a study on baseline electricity demand and productive use at the village level in four Indian states (Bihar, Uttar Pradesh, Odisha, and Rajasthan), with a focus on the rural population. Based on the findings, rural households have high grid-electrification coverage and adoption, with the electric grid serving as many people's major source of power and illumination. Rural firms have affordability challenges due to high interconnection fees and limited access to reliable power sources [11]. To address this, non-grid alternatives like solar residential systems, rechargeable batteries, micro-grids, and diesel-fuelled generators play an essential role in rural power distribution. Non-grid energies are used by 16% of homes and 40% of industries. Given India's favourable climate, many people opt to use solar electricity.

The issues with energy production and use are linked not just to climate change but also to sustainability issues that include polluted air, the formation of acid rain, diminished ozone layer, degradation of forests, and toxic emissions [12]. To mitigate these repercussions, several possible options have become apparent, notably preserving energy by means of improved energy usage, decreasing dependence on fossil fuels, and more investment in ecologically friendly energy suppliers [13].

In addition, the IPCC made three major remarks [14]: (i) Humans have an apparent effect on the climate cycle; (ii) CO2 concentrations in the atmosphere have reached a record high in the past 800,000 years; and (iii) CO2 concentrations produce the vast majority of the overall radiative forcing in the atmosphere. The past researches demonstrate the clear link between global temperature and planetary carbon dioxide concentration, as well as how upcoming global temperature rises may be mitigated by lowering CO2 emissions. Similar investigations found that climate change will have far more detrimental than favourable consequences in the years to come. Key concerns include food instability, fresh water availability, receding waters, adverse weather conditions, and destruction of marine and coastal habitats. Evidently, the continued existence of ecosystems that promote routine activities demands a reduction of subsequent ambient CO2 emissions. Therefore, solar energy, a low-carbon energy source, is essential for reducing the effects of changes in the climate [15].

Solar energy, among other renewable energy sources, is a potential and widely accessible energy option for addressing prospective energy crisis challenges. Because of the tremendous need for energy, the solar sector is rapidly expanding throughout the world, due to the fact that the primary energy source, fossil fuel, is finite and other sources are costly [16–17]. The planet receives solar energy in a variety of forms, including light and heat. A significant percentage of this energy is lost during transit due to cloud adsorption, reflection, and dispersion. Nonetheless, several studies have found that solar energy is a low-cost and abundant natural resource capable of meeting global energy demands.

PV cells are currently widely employed due to their economic viability and minimal carbon impact. Because of the environmentally beneficial nature of electricity generation, it is important to investigate various types of photovoltaics [18]. The net environmental impact for silicon cells, both monocrystalline and polycrystalline, is between 20 and 25 g of CO2 equivalent per unit (kWh) of power. Solar PV also has a far shorter return on investment. It takes around 1 year and 2 years for pc-Si and mc-Si, accordingly [19].

2. The sun's energy

Solar energy is the most plentiful renewable energy source, with the sun emitting 0.38×10^{24} kW and the earth receiving around 180×10^{12} kW [20]. Moreover, it is a potential source of energy because of its non-exhaustibility and higher efficiency compared to other sources. Solar radiation dispersion and concentration/intensity play a crucial role in the efficiency of the solar PV sector. These two characteristics are widely varied among countries. Specific yield (kWh/kWp) is a popular performance indicator for solar power plants of any dimension. It is used for assessing locations, analysing layouts, and evaluating the overall condition of a PV array. **Figure 1** depicts the global peaks for solar power prospective in terms of kWh/kWp. It is determined by a few parameters, including location, weather, module position, module preference, and balance of system efficiency, BoS. It is observed from **Figure 1** that Asian

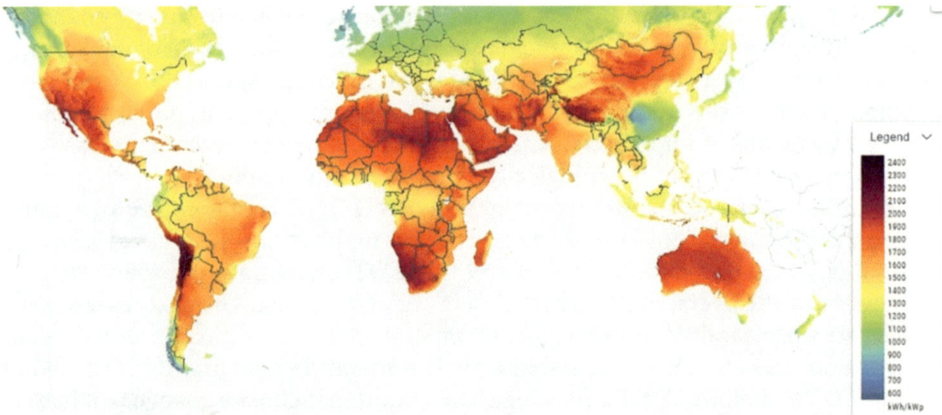

Figure 1.
Global solar photovoltaic power potential [21].

countries possess the biggest potential for solar radiation due to their prolonged sunlight duration throughout the year, surpassing other tropical nations.

A solar cell, also known as a photovoltaic cell, is a semiconductor-based device that quickly converts sunlight into energy. Light hitting a cell generates voltage and current, resulting in electrical power. To do this, two criteria must be adhered to: (i) a material capable of raising electrons to higher energy levels through photon absorption, and (ii) the ability to move excited electrons from the PV cell to the external load.

The electron gets back to the PV cell after discharging energy into the peripheral circuit. There are several materials available to meet energy conversion requirements. In practical scenarios, the production of electricity from sunlight relies solely on semiconducting materials that create PN junctions. PV units are often installed on building facades and roofs.

Solar photovoltaic systems operate similarly to diodes, utilising the PV effect. It is a mechanism for transferring heat energy from the sun into electrical energy.

They emit no CO2 and are thus environmentally beneficial. These materials emit photons, which facilitate electron mobility.

Figure 2 depicts the major components of a solar PV module. Solar cells are the fundamental components of solar panels. Several thousand of such cells combine to make a solar panel. They are strung together to form photovoltaic panels, which require soldering, encapsulation, installation on a metal frame, testing, and so on. The yield of a solar panel is proportional to that of solar cells. The price and efficiency of solar cells impact the total output of a solar panel. The efficiency of solar cells also varies depending on the materials employed. There are two primary varieties of solar cells: monocrystalline and polycrystalline.

An additional crucial part of a solar panel is solar glass. It is the solar panel's outermost layer, and it must be robust and polished for the panel to work well. The primary function of solar glass is to shield solar cells from inclement weather, dirt, and debris. Tempered glass with a thickness of 3–4 mm is ideal. The EVA sheet, also known as 'ethylene vinyl acetate', is a very transparent (plastic) covering that is employed to enclose cells. It creates a fused layer on the surface of the cells to keep them intact. It must be both sturdy and resilient to tolerate severe temperatures and humidity. The backsheet is the panel's last layer, offering mechanical safety as well as insulation against electricity. It is primarily a layer of safeguarding.

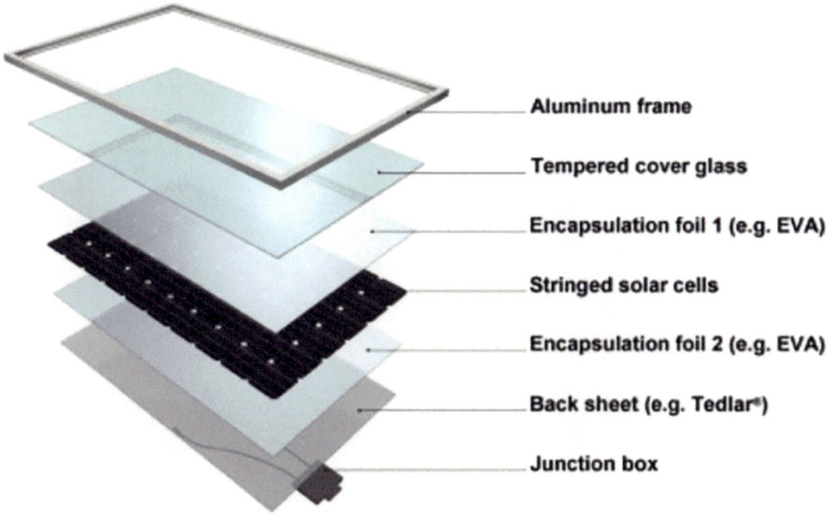

Figure 2.
Principal elements of a photovoltaic module [22].

The aluminium framing is also a vital element since it offers structural support for the panel. It is advisable to use a frame composed of a durable yet thin material. It needs to be sufficiently firm to endure harsh circumstances such as strong winds and external pressures. It is normally available in two finishes: silver and anodised black. A junction box is attached to the rear of the panel. It is the centre place where wires connect to the panels. Interconnectors enable solar panels to communicate with each other. They require being very weatherproof and provide reliable connections. Silicon is the most popular solar panel adhesive. Silicon forms strong connections and is immune to compounds, water, and environmental factors. Hence, silicon glue is employed for solar panels. This is perhaps a very popular semiconductor substance.

3. PV cell operation

When sunshine (photons) reaches the outermost layer of a PV cell, it permeates the cell's semiconductor material (typically silicon). The energy of captured photons is transmitted to electrons in the semiconductor material, enabling them to break away from their atomic connections and form pairs of electron-holes. Electrons are negatively charged and may travel easily, whereas holes are charged positively. The freshly formed electron-hole pairs get separated by the electric field built into the PV cell (made by the junction of multiple semiconductor layers). The separated electrons are gathered by metal electrodes on the cell's surface, resulting in electric current. This current can be extracted and used externally. Whenever an external electric load (which could be an illumination source or a battery) is connected to the PV cell, the passage of electrons via the load produces electrical power, which may then be utilised for powering multiple appliances or preserved.

in batteries for future use. The photovoltaic cell uses this mechanism to continue producing electricity for as long as sunlight is present, making it an energy source that is both renewable and sustainable. The operation of a solar photovoltaic cell is explained in **Figure 3(a)** and **(b)**.

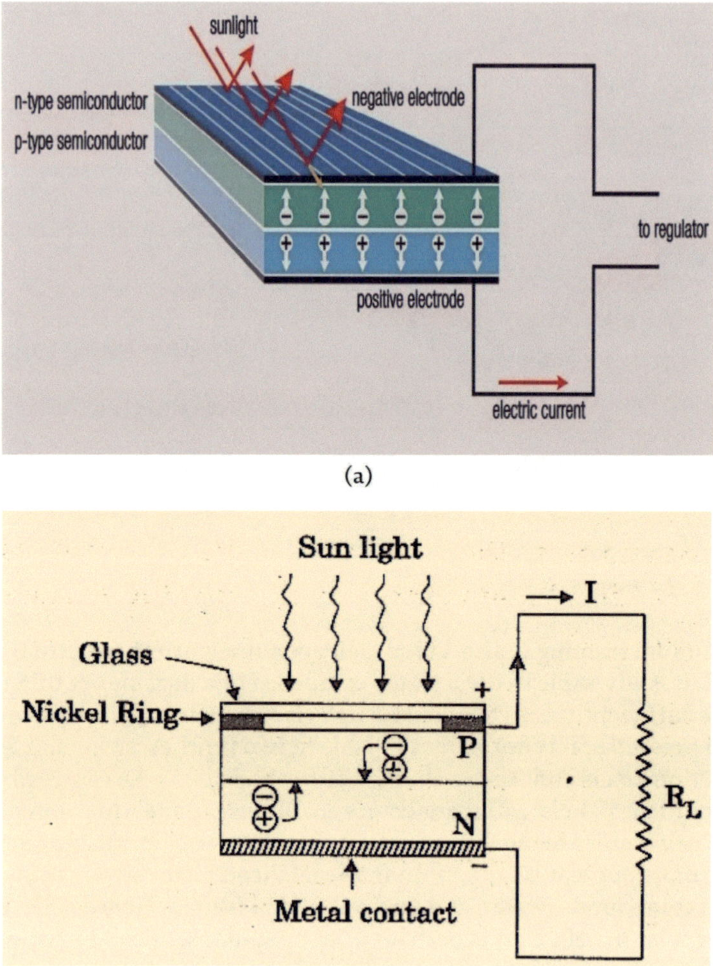

(a)

(b)

Figure 3.
(a). Working principle of solar PV cell [23]. (b). P-N junction: Solar PV cell.

4. Materials and solar cell performance

The Photo-Conversion Efficiency (PCE), or simply the conversion efficiency, of a solar cell implies its capacity of converting sunshine into useful electrical power. It is a critical metric that reflects how well the solar cell turns sunlight into energy. The current density of a solar cell is measured as the voltage across the device is biased with a changing load during exposure to sunlight (**Figure 4**). The PCE is related to the short-circuit current density (Jsc) and the open-circuit voltage (Voc) and is given in Eq. (1) below.

$$PCE = \frac{J_{sc} V_{oc} FF}{P_{solar}} \tag{1}$$

The terms FF and P_{solar} relate to the fill factor and incident electrical power from solar irradiation, accordingly [13]. The highest voltage that can be attained by a solar

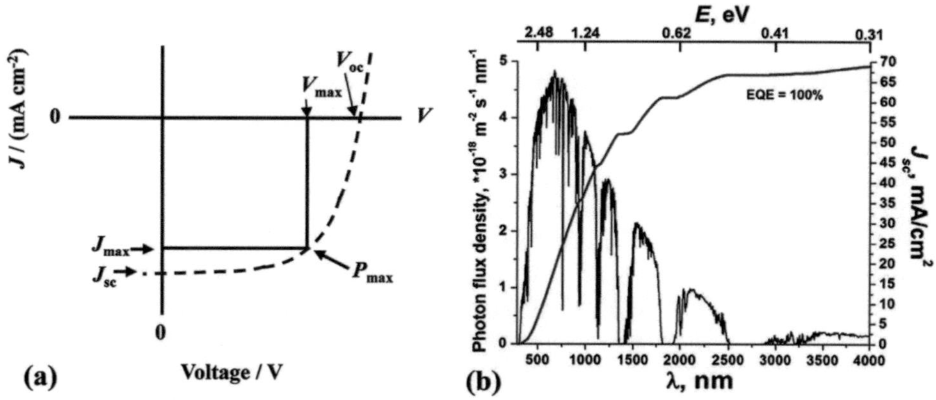

Figure 4.
Solar cell efficiency and solar radiation. (a) J-V Characteristics of a solar cell; (b) The photon-flux characteristics [15].

cell having its contacts open, or the voltage that can be acquired from a cell with zero current flowing in it, is termed the open-circuit voltage, V_{oc}. Short-circuit current is the current that occurs when the two terminals of a cell are shorted together. The voltage at this point is zero. This current for the specified area is referred to as J_{sc}. The quantity of FF is the ratio of P_{max} (product of J_{max} and V_{max}) to the product of J_{sc} and V_{oc}, and it measures the rectangular shape of the J-V characteristic.

The price and efficacy of manufacturing a solar cell are determined by the material used. An extensive study has been conducted in this domain to determine the most affordable and effective material for the construction of solar cells.

The requirements for high-quality materials for PV solar cells comprise the following [24]:

- band gap between 1.1 and 1.7 eV
- direct band structure
- ease of access
- non-toxicity
- high photovoltaic efficiency

A major issue in solar cell research is the finding of ways to obtain the best feasible efficiency while minimising manufacturing expenses. Solar cell efficiency may be increased by employing efficient methods to decrease the cell's inherent losses. There are three primary categories of losses: optical, quantum, and electrical. Each category has a distinct source. Lowering losses of any type necessitates the use of multiple, typically sophisticated, cell and solar panel manufacturing technologies. The renowned Shockley-Queisser (SQ) limit, which considers the balance between photogeneration and radiative recombination, establishes an upper efficiency limit on the market-available technology [25].

5. Generations of photovoltaic cells

Developments in manufacturing techniques and materials have played an essential part in the growth of PV cells. The production process for solar cells involves various techniques, including modifications to materials with varying photoelectric conversion capacities in the cell sections. Photovoltaic technologies may be classified into four primary generations as a result of the introduction of several non-traditional ways of production to manufacture functional solar cells.

First-generation solar cells are built on crystalline Si (c-Si), but second-generation solar cells are centred on thin film technology, and this frequently uses vapour-deposited semiconductors. Third-generation solar cells, on the other hand, are solution-processed solar cells made of organic compounds, hybrids, and inorganic semiconductors, as well as nanostructured substances. Fourth-generation solar cells are the latest type of photovoltaic devices designed to circumvent the constraints of existing solar cells by providing better efficiency, reduced cost, and enhanced usability. **Figure 5(a)** is

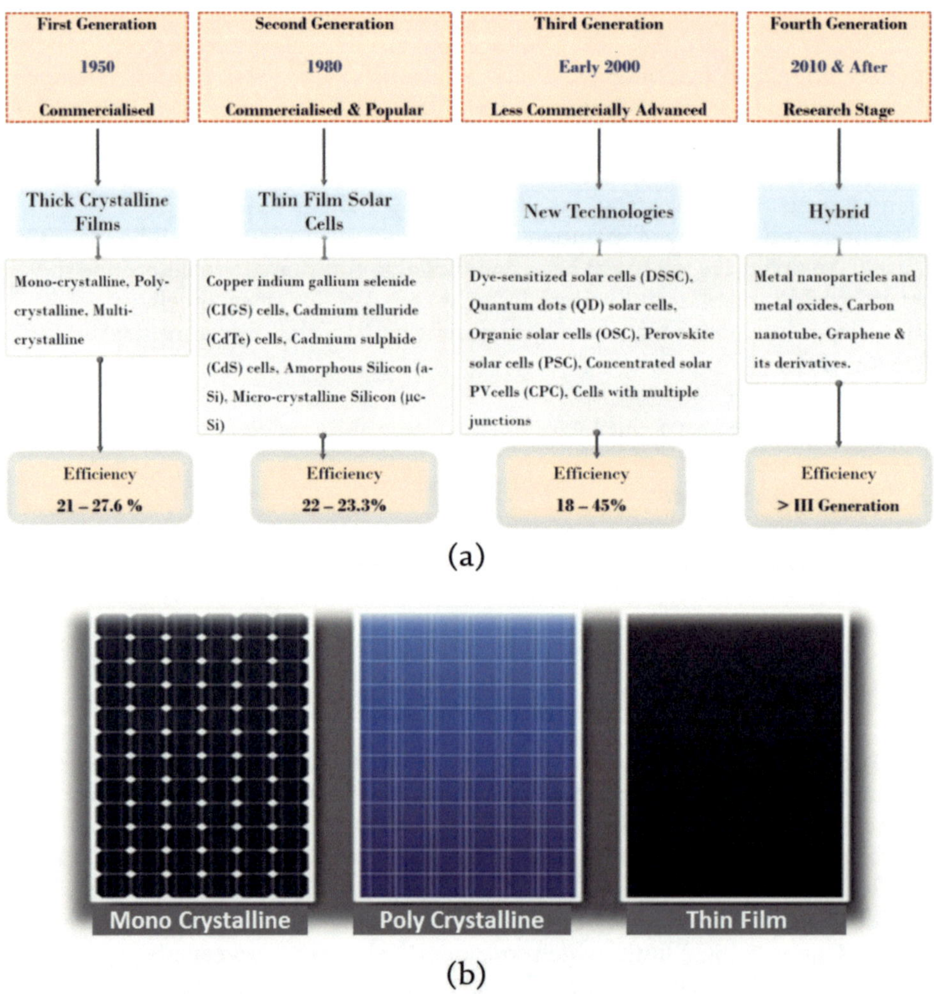

Figure 5.
(a). Various generations of PV cell technologies. (b). Commercial types of PV panels.

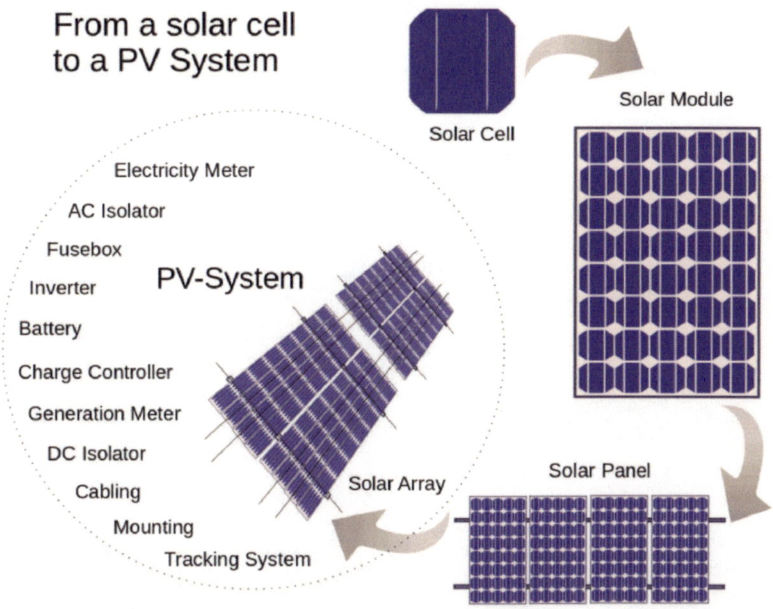

Figure 6.
A PV system from a solar cell.

used to describe the different generations of PV cells. Numerous studies and reports have investigated the various materials employed in first- and second-generation PV, but this study focuses on emerging generation materials. **Figure 5(b)** depicts the most prevalent types of PV panels available on the market.

The voltage of a solar cell, solar module, solar panel, and solar PV system varies according to the number of cells, how they are linked, and the particular type of system (**Figure 6**). A single solar cell generates an open-circuit voltage of around 0.5–0.6 Volts. The voltage drops to around 0.46 Volts while under load. A solar panel is a structure made up of several solar cells and normally generates 12–24 Volts. In a solar panel array, the number of modules linked in series determines the voltage output. By connecting numerous solar panels in series or parallel, respectively, a solar PV system's voltage and current output may be enhanced. Joining panels in series increases the voltage, while assembling them in parallel increases the current.

A solar PV system is made up of multiple components, as shown in **Figure 6**, to ensure the system's safety and to suit varied demands at an affordable price. The purpose of the charge controller is to control the current flowing between the battery bank and the panels, while the inverter transforms DC electricity into AC for usage in the house or grid connection. For usage in low- or no-light conditions, the battery bank stores solar energy.

6. Structures of organic solar cells

Organic solar cells used in the latest PV generations feature photo-active layers that are typically made up of a semiconducting polymer and fullerene derivatives. This class includes Bulk Hetero-Junctions (BHJs), which are the interfaces of two distinct semiconducting materials [24]. This sort of cell offers a conversion efficiency

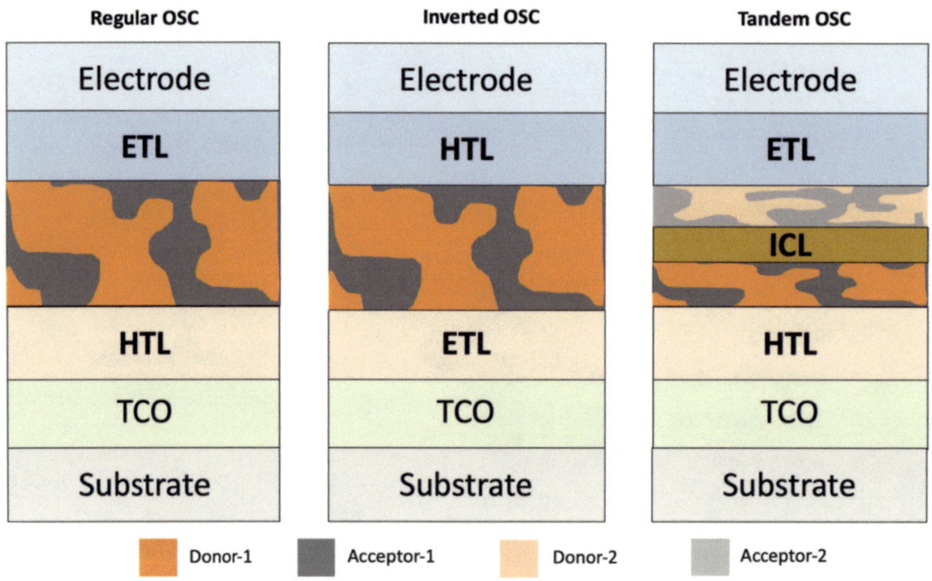

Figure 7.
Structure of organic solar cell showing bulk hetero-junctions [16].

of around 18%. Organic solar panels have several important benefits over inorganic components, including inexpensive material costs, soft, durable, and adjustable absorption properties, adaptability, and the ability to be manufactured utilising roll-to-roll or sheet-to-sheet printing processes. Yet certain organic compounds have stability problems due to photochemical breakdown. Organic solar cells may be made translucent or various colours however, this reduces their total efficiency.

An organic solar cell is constructed using a Transparent Conductive Oxide (TCO) electrode material, such as Indium Tin Oxide (ITO) substrate. **Figure 7** illustrates the three varieties of organic solar cells referenced in most of the scientific literature. In the regular BHJ arrangement, ITO is plated with a thin film of Hole Transport Layer (HTL); over this HTL, a mix of donor and acceptor is coated, and a small Electron Transport Layer (ETL) with a low work-function metal cathode is typically vapour-deposited. The HTL and ETL switch places in relation to the standard layout in the "inverted" version because of the inverted polarity of charge collection. The BHJ mix is covered with a thin layer of high work-function metal anode and HTL.

In any case, both of the aforementioned single-junction arrangements can only accomplish an ultimate efficiency of 14–16% in the best-case scenario owing to either the donor/acceptor blend's lack of a wide absorption range or restricted charge carrier accessibility, which necessitates a thin (80–120 nm) blend for successful charge collection.

To overcome these drawbacks, at least two sub-cells having complementary and/or congruent light absorption characteristics are joined by a thin interconnecting layer (ICL) to form a tandem structure.

7. Structures of perovskite solar cells

Perovskite solar cells are modern third-generation solar cells that look to have a very strong probability of aiding in large-scale generation of solar power because of their substantial PCE and compliance with expandable techniques. A perovskite solar cell is a several-layered device made up of a Transparent Conducting Oxide (TCO) front electrode (typically based on FTO for Fluorine Tin Oxide), a light-absorbing perovskite layer placed between the N-type Electron Transport Layer (ETL) and the P-type Hole Transport Layer (HTL), along with a metal back electrode. These cells may be made in either a conventional n-i-p or an inverted p-i-n arrangement, with the first being split into mesoporous and planar structures, while the second type is typically planar. ETL and HTL are chosen depending on the structure's energy-band coherence and the processing solvents' suitability.

Currently, the most difficult challenge with perovskite solar cells is their stability over time, which must be resolved before they can be used in real-world applications. As we know, the durability of perovskite solar cells under harsh environments, such as temperature treatment, light irradiation, humidity, and so on, seems to be a barrier preventing their wider adoption. Humidity has been identified as one of the probable reasons for perovskite deterioration. The commercial use of perovskite solar cells necessitates substantial investigation and invention of novel perovskite materials that are not only highly efficient in photoelectric conversions but also harmless and resilient.

8. Fourth-generation solar cells

Because they integrate the affordability and adaptability of polymer thin films plus the stability of organic nanostructures like metal nanoparticles and oxides, carbon nanotubes, graphene, and its derivatives, fourth-generation solar power cells are sometimes referred to as hybrid or composite inorganic cells. These devices, frequently mentioned as "nanophotovoltaics," might represent the optimistic potential of photovoltaics [25, 26].

Graphene-based solar energy cells: The fourth generation delivers exceptional cost-effectiveness and versatility by combining thin polymer films and metal nanoparticles with diverse metal oxides, carbon nanotubes, graphene, and its derivatives [27]. Graphene received special attention since it is seen as a potential nanomaterial. Because of its unique features, including high carrier mobility, low resistivity plus transmittance, and two-dimensional lattice packaging, graphene-based substances are being evaluated for application in solar cells rather than traditional materials. Still, the primary factor for attaining acceptable device performance in practical uses is the production of graphene materials with suitable geometry and characteristics.

Graphene-polymer-based solar cells: They have gained prominence in the industry, and such polymer components may be organic and advantageous owing to their easy manufacturing process, affordable price, and ease of tuning. As a hybrid material composed of organic and inorganic components, graphene has been employed in applications such as coating, stacking, and thermal annealing of electrodes. Graphene and semiconductor films have a lower Fermi level, making them ideal for charge injection uses as well as robust energy properties. Graphene may be mixed with polymers to form a hybrid material with an estimated bandgap of 3.6 V, preventing electron passage from the cathode to the electrode [28]. Graphene sheets with a thickness of 2 nm are said to produce the optimal performances with regard to both electrical resistance and electron transfer.

Other types of graphene-based solar energy cells: Carbon allotropes that consist of graphene and carbon nanotubes, if combined with quantum dots, may generate effective solar cells, and the graphene-quantum dot combination offers various possible uses. Tandem solar cells are made up of more than two sub-cells, known as multi-junction solar cells. In accordance with study findings, a single solar cell is around 40% effective; however, tandem solar cells constructed from graphene oxides are as much as 86% effective due to their more expansive subcell combo.

9. Comparison among various PV generation technologies

Solar energy is a clean and sustainable energy source that contributes to reaching net zero carbon emissions. However, the conversion efficiency and technology utilised to convert sunlight into energy are critical factors in determining the overall system cost. Currently, the power tariffs for fossil fuels are lower than those for solar energy. Continuous research and development have decreased system costs while increasing both lifespan and conversion efficiency.

However, these benefits are dependent on the type of material utilised, which determines the final cost. The location, direction, and application are other important considerations for achieving optimal efficiency at the lowest possible cost. **Table 1** provides an overall comparison of several PV-generating technologies.

Figures 8 and **9** illustrate the maximum efficiency and life lifetime of PV cells up to the third and second generations, respectively, while **Table 1** provides an overall comparison of the different PV generation cells.

10. Concluding remarks

To comprehend the development of PV cells, their fundamental constraints, and their implications for new energy, a variety of reports and models, including IEA reports, Panel Threshold Regression Models, SPI Study, and IPCC Guidelines, have been researched. Solar energy constitutes one of the most in-demand green power sources. Electricity generation utilising solar technology not simply contributes to meeting increasing needs for energy, yet additionally assists in reducing worldwide climate change by lowering reliance on fossil fuels. The viability of new future-oriented solar cells is improving as a result of researchers and scientists' attempts to produce novel materials, notably nanomaterials and nanotechnology.

Solar PV panels are made of a number of semiconductor materials, including silicon. The output or result of a solar panel is mostly determined by the short-circuit current and open-circuit voltage. They in turn give enough maximum power. The band gap energy is impacted by all of the characteristics mentioned above. In reality, this energy determines the fill factor and efficiency of the solar panel. Thus, choosing the right material is critical to the functionality of PV cells. Despite the fact that fourth-generation PV cells are becoming increasingly prevalent, crystalline forms continue to be more efficient. Monocrystalline structures are more efficient and have a smaller environmental impact than polycrystalline structures. Yet, the multi-layer structure and multi-spectrum capabilities of new PV cells are economically feasible, but further study is required to provide higher efficiencies with fewer problems.

While first- and second-generation photovoltaic technologies (monocrystalline and polycrystalline silicon) are still widely used due to their dependability and market maturity, significant advances in third- and fourth-generation solar cells

Generation	Materials	Band gap (eV)	Efficiency (%)	Life span (Years)	Benefits	Limitations
First (1950)	*Monocrystalline silicon (m-si)*	1.1	15–24	25	Reliability, great performance, and a substantial life expectancy.	High production costs, increased temperature sensitiveness, absorption issues, and material loss.
	Polycrystalline silicon (p-si)	1.7	10–18	14	The manufacturing process is easy, lucrative, reduces silicon waste, and has greater absorption than m-si.	Reduced efficiency, increased temperature susceptibility.
	Solar cells based on GaAs	1.43	28–30	18	Excellent stability, decreased temperature sensitivities, superior absorption to m-si, and high performance.	Very costly.
Second (1980)	*Amorphous silicon (a-si)*	1.7	5–12	15	Less costly, widely available, non-toxic, with a high absorption index.	Reduced efficiency, difficulties identifying dopant materials, and short minority carrier lifetimes.
	Cadium telluride/cadium sulfide (CdTe/CdS)	1.45	15–16	20	Good absorption rate requires less material for manufacturing.	Decreased efficiency, very poisonous Cd, limited Te, and higher temperature sensitivity.
	Copper indium gallium selenide (CIGS)	1.7	20	12	Less material is needed for its production.	Very expensive, unstable, more dependent on temperature, and extremely unreliable.

Generation	Materials	Band gap (eV)	Efficiency (%)	Life span (Years)	Benefits	Limitations
Third (early 2000)	*Dye-sensitised photovoltaic cells*	—	5–20	—	Reduced cost, dim light and broader angle operation, reduced internal temperature operation, ruggedness, and longer lifetime.	Issues with temperature stability, toxic and volatile chemicals.
	Quantum dots	—	*11–17*	—	*Low manufacturing costs, low energy use.*	*Extremely toxic in nature.*
	Organic and polymeric photovoltaic cells	—	*9–11*	—	*Low processing costs, lower weight, flexibility, and thermal stability.*	*Inadequate productivity.*
	Perovskite solar cells	—	*21*	—	*Cheap cost and simple construction, light weight, flexibility, high efficiency, cheap production costs.*	*Unstable.*
	Multi-junction solar cells	—	*36%*	—	*Excellent performance.*	*Sophisticated and pricey.*
Fourth (Research/ Upcoming)	*Nanomaterials, metal oxides, carbon nanotubes, and other technologies are used to overcome previous generations' barriers and collect maximum energy at a cheap cost and with stability. Efficiency is determined by the materials and procedures utilised, and it varies depending on the application.*					

Table 1.
Comparison among various PV solar cell technologies.

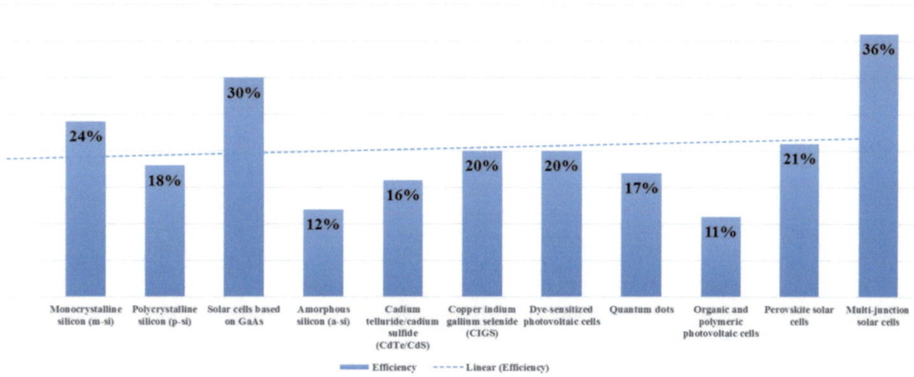

Figure 8.
Maximum efficiency of various PV materials.

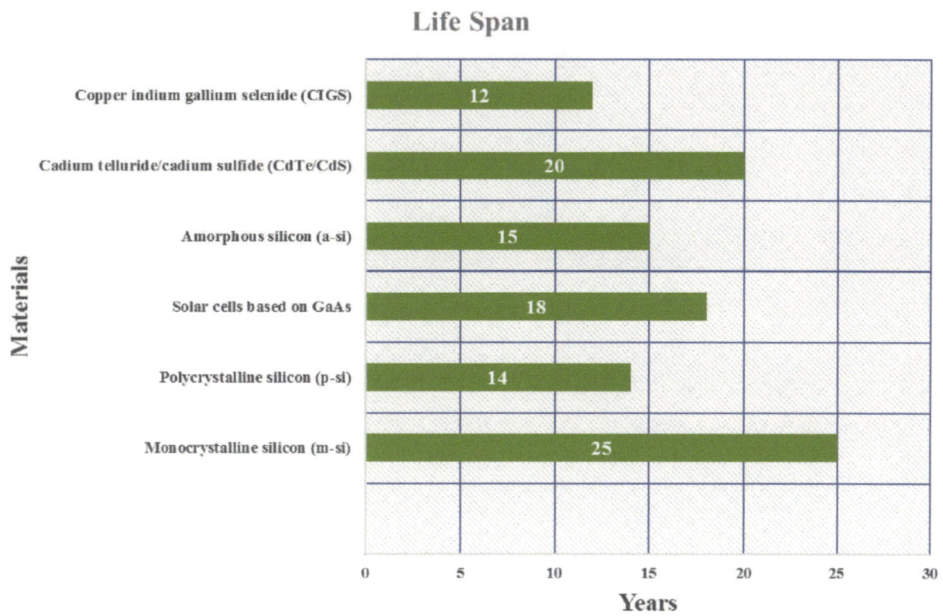

Figure 9.
Life span of a few PV materials.

(e.g., organic, perovskite, and nanomaterial-based cells) are expected to transform the industry. These modern technologies are challenging the limits of efficiency, adaptability, and cost-effectiveness. Although there are still issues with their long-term stability and scalability, perovskite cells, with their high power conversion efficiency, and nanomaterials like graphene have the potential to improve the performance of solar panels.

By integrating conventional materials with nanotechnology, fourth-generation PV technologies are emerging that provide the most potential path to overcome the constraints of existing solar technologies, substantially improving durability and efficiency while lowering prices. This will be crucial in dealing with the global energy problem and promoting sustainable development.

Conflict of interest

The authors declare no conflict of interest.

Author details

Sivasami Kolandasamy[1] and Thangalakshmi Sivalingam[2*]

1 Indian Maritime University, Kolkata Campus, India

2 Indian Maritime University, Mumbai Port Campus, India

*Address all correspondence to: thangalakshmiprakash@yahoo.com

References

[1] IEA. Key World Energy Statistics 2020. Paris: IEA; 2020. Available from: https://www.iea.org/reports/key-world-energy-statistics-2020, Licence: CC BY 4.0

[2] Ratna R, Chaudhary V. Energy consumption in urban India: Usage and ignorance. Urban Energy Systems: Modeling and Simulation for Smart Cities. 2023;**22**:65-80

[3] Chaudhari A, Mulay P. Algorithmic analysis of intelligent electricity meter data for reduction of energy consumption and carbon emission. The Electricity Journal. 2019;**32**(10):106674

[4] Shahee A, Abdoos M, Aslani A, Zahedi R. Reducing the energy consumption of buildings by implementing insulation scenarios and using renewable energies. Energy Informatics. 2024;**7**(1):18

[5] Noviello L. What Is the Carbon Footprint and How Is it Calculated? Enel Green Power; 2023. Available from: https://www.enelgreenpower.com/learning-hub/energy-transition/carbon-footprint

[6] Haugland T, United Nations Economic Commission for Europe, & Global Methane Initiative. Best Practice Guidance for Effective Methane Management in the Oil and Gas Sector. In: De La Flor Garcia F, Diamond J, Drabik M, Foster S, Hight C, Hooper D, et al., editors. 2019. Available from: https://unece.org/fileadmin/DAM/energy/images/CMM/CMM_CE/BPG_Methane_final_draft_190912.pdf

[7] Li R, Wang Q. Does renewable energy reduce per capita carbon emissions and per capita ecological footprint? New evidence from 130 countries. Energy Strategy Reviews. 2023;**49**:101121

[8] Ehigiamusoe KU, Dogan E. The role of interaction effect between renewable energy consumption and real income in carbon emissions: Evidence from low-income countries. Renewable and Sustainable Energy Reviews. 2022;**154**:111883

[9] Achuo ED, Miamo CW, Nchofoung TN. Energy consumption and environmental sustainability: What lessons for posterity? Energy Reports. 2022;**8**:12491-12502

[10] Adams S, Nsiah C. Reducing carbon dioxide emissions; does renewable energy matter? Science of the Total Environment. 2019;**693**:133288

[11] Holttinen H, Jørgensen BH. IEA Wind TCP Annual Report 2022 Executive Summary. IEA; 2023

[12] Kannan N, Vakeesan D. Solar energy for future world—A review. Renewable and Sustainable Energy Reviews. 2016;**62**:1092-1105

[13] Chernyakhovskiy I, Joshi M, Gadzanku S, Inskeep S, Rose A. Opportunities for Renewable Energy, Storage, Vehicle Electrification, and Demand Response in Rajasthan's Power Sector. Golden, CO (United States): National Renewable Energy Lab. (NREL); 2022

[14] Stocker TF, Qin D, Plattner GK, Alexander LV, Allen SK, Bindoff NL, et al. Technical summary. In: Climate Change 2013: The Physical Science Basis. Contribution of Working Group I to the Fifth Assessment Report of the Intergovernmental Panel on Climate Change. Cambridge University Press; 2013. pp. 33-115

[15] Yan J, Saunders BR. Third-generation solar cells: A review and comparison of

polymer: Fullerene, hybrid polymer and perovskite solar cells. RSC Advances. 2014;**4**(82):43286-43314

[16] Miyake S, Teske S, Rispler J, Feenstra M. Solar and wind energy potential under land-resource constrained conditions in the Group of Twenty (G20). Renewable and Sustainable Energy Reviews. 2024;**202**:114622

[17] Mao D, Yang S, Ma L, Ma W, Yu Z, Xi F, et al. Overview of life cycle assessment of recycling end-of-life photovoltaic panels: A case study of crystalline silicon photovoltaic panels. Journal of Cleaner Production. 2023;**434**:140320

[18] Thangalakshmi S, Ganeshram V, Sivasami K, Sathiyasekar K. Effect of panel materials on the performance of solar cells. PalArch's Journal of Archaeology of Egypt/Egyptology. 2020;**17**(7):5329-5340

[19] Müller A, Friedrich L, Reichel C, Herceg S, Mittag M, Neuhaus DH. A comparative life cycle assessment of silicon PV modules: Impact of module design, manufacturing location and inventory. Solar Energy Materials and Solar Cells. 2021;**230**:111277

[20] Panwar NL, Kaushik SC, Kothari S. Role of renewable energy sources in environmental protection: A review. Renewable and Sustainable Energy Reviews. 2011;**15**(3):1513-1524

[21] Koster G, van Sark W, Ricker B. Solar potential for social benefit: Maps to sustainably address energy poverty utilizing open spatial data in data poor settings. Energy for Sustainable Development. 2024;**80**:101453. ISSN 0973-0826. DOI: 10.1016/j.esd.2024.101453. Available from: https://www.sciencedirect.com/science/article/pii/S0973082624000796

[22] Chandra N. What are the Main Components of a Solar Panel? LOOM SOLAR; 30 Jul 2024b. Available from: https://www.loomsolar.com/blogs/collections/solar-panel-components?srsltid=AfmBOoqldO5KFe5iq9qk38OWvbHo-anjC77COkFxxK_0MblI4u6k3sYud

[23] Team C. How Is A Solar Cell Made: Construction, Working & Power Output. (n.d.). Available from: https://solarsquare.in/blog/solar-cell-construction/

[24] Pastuszak J, Węgierek P. Photovoltaic cell generations and current research directions for their development. Materials. 2022;**15**(16):5542

[25] Tayebjee MJ, McCamey DR, Schmidt TW. Beyond Shockley–Queisser: Molecular approaches to high-efficiency photovoltaics. The Journal of Physical Chemistry Letters. 2015;**6**(12):2367-2378

[26] Dos Santos DR, Julien-Vergonjanne A, Dkhil SB, Parmentier M, Combeau P, Sahuguède S, et al. Toward indoor simulations of OPV cells for visible light communication and energy harvesting. IEEE Access. 2024;**12**:41027-41041

[27] Das S, Pandey D, Thomas J, Roy T. The role of graphene and other 2D materials in solar photovoltaics. Advanced Materials. 2019;**31**(1):1802722

[28] Aamir Iqbal M, Malik M, Shahid W, Zaheer Ud Din S, Anwar N, Ikram M, et al. Materials for Photovoltaics: Overview, Generations, Recent Advancements and Future Prospects. IntechOpen; 2022. DOI: 10.5772/intechopen.101449

Chapter 4

Perovskite-Structured Methylammonium-Tin Halides for Optoelectronic Applications

Nagaraj Gopinathan and Syed Abdul Gafoor Sathik Basha

Abstract

Organic-inorganic halide perovskite-structured (OIHPS) materials emerge as promise next-generation optoelectronic materials. In particular, low-toxic methylammonium-tin halides ($CH_3NH_3SnX_3$; X = I, Br, Cl) exhibit strong electrical and tunable optical properties. The properties of a perovskite material strongly depend on the structure, morphology, selection of solvent, and composition of the films. The state-of-the-art research efforts focused on $CH_3NH_3SnX_3$ (X = I, Br, Cl) perovskite, and these applications are thoroughly reviewed in this chapter. Furthermore, the enhancement of structural, morphology, optical, electrical, and thermal properties of $CH_3NH_3SnX_3$ (X = I, Br, Cl) perovskites was also discussed. We conclude by discussing the limitations and potential advancements of these $CH_3NH_3SnX_3$ (X = I, Br, and Cl) perovskite materials.

Keywords: OIHPS, $CH_3NH_3SnX_3$, tunable properties, solar PV, optoelectronics

1. Introduction

Of late, organic-inorganic halide perovskite-structured (OIHPS) have recently attracted huge attention in the fields of solar cells [1], light-emitting diodes (LEDs) [2], photodetectors [3] and sensors [4] due to their greater light-absorbing and emitting properties. It can be synthesized at a minimal cost with a simple low-temperature solution procedure [5]. The OIHPs have an ABX_3 structure, where A is a larger organic cation occupied with 12 X anions in the body-centered positions (i.e., $CH_3NH_3^+$, $HC(NH_2)_2^{+}$, or Cs^+). Typically, B is smaller divalent metal ions occupied with six X anions in the cube corners (i.e., Pb^{2+}, Sn^{2+}, Ge^{2+}, or Sr^{2+}). For X is mixture of monovalent anions in the face-centered positions (i.e., Cl, Br, or I) is represented in **Figure 1**. The obtained crystal structure of perovskites is depended upon the Goldschmidt tolerance factor (t). It is formula is given below:

$$t = \frac{rA + rX}{\sqrt{2}(rB + rX)} \tag{1}$$

IntechOpen

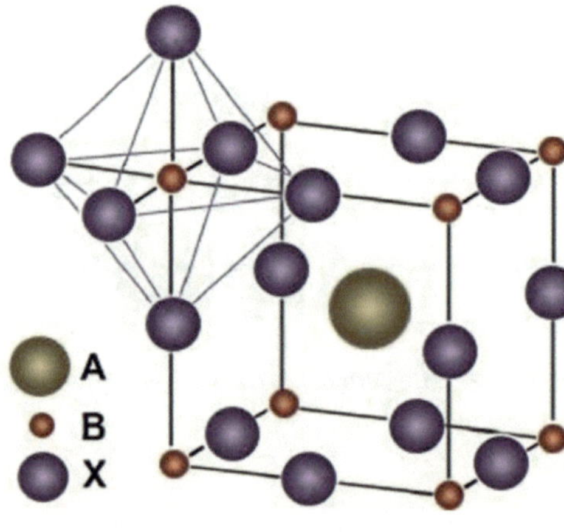

Figure 1.
ABX_3 perovskite structure, A represents the Cs^+, B represents the Pb^{2+}, Sn^{2+}, Ge^{2+}, or Sr^{2+}, and X represents the Br^-, Cl^-, I^- ions (reproduced with permission of Ref. [6]).

For *rA, rB*, and *rX* are ionic radii of each A, B, and X ions. Here, t > 1 indicates the structure is tetragonal or hexagonal. If t = 0.9 to 1 in the range structure should be cubic structure. If t < 0.9, it is the orthorhombic or rhombohedral structures.

The octahedral factor (μ) is determined by using formula is given below:

$$\mu = \frac{rB}{rX} \tag{2}$$

where *rA* and *rX* are ionic radii of B and X ions.

The most studied and efficient OIHPS contains lead as a metal ion on the B-site. Lately, many researchers have addressed OIHPS instability and toxicity of lead [7]. This is hidden from large-scale commercial implementation. Various intrinsic and extrinsic variables contribute to the instability of hybrid perovskite systems, and various synthesis strategies have been proposed to build a stable perovskite structure, i.e., ligands-assisted synthesis, encapsulation, and dual-phase synthesis [7]. In replacing toxic Pb^{2+}, various approaches have been proposed so far, including Sn^{2+}, Ge^{2+}, Bi^{3+}, and Sb^{3+}, as B-site materials [8]. According to our knowledge, tin (Sn^{2+}) is a suitable material and an environmentally friendly alternative to Pb^{2+}. However, utilizing lead in these compounds poses toxicity concerns throughout the device lifecycle. To circumvent this issue, promising alternatives such as tin, belonging to group 14 of the periodic table, offer comparable physical and chemical properties. While there are similarities between lead-based and tin-based perovskite systems, differences arise from the distinct ionic radii of Sn^{2+} and Pb^{2+} ions. Consequently, the chemical bonding in Sn systems becomes more covalent due to greater orbital overlap in shorter Sn-X bonds compared to Pb-X bonds, affecting semiconductor properties. Additionally, relativistic effects are stronger in Pb, leading to a larger stabilization of its $6s^2$ electrons compared to Sn $5s^2$ electrons, resulting in a narrower band gap and broadened bandwidths in Sn systems [9]. Various chemical methods, including halide mixing and crystal growth control, along with different processing approaches, have been developed to optimize the properties of perovskite absorbers and enhance device performance [10]. Perovskite-structured methylammonium-tin halides for optoelectronic applications and their advantages over other materials utilized for optoelectronic device fabrication [11].

2. Perovskite synthesis methods

There are many proposed methods for the synthesis of organic-inorganic halide perovskite-structured (OIHPS) by many groups in the literature, i.e., precipitation method, ball milling, vapor deposition, dip coating, spin coating, and top-seeded solution method are most commonly used techniques.

2.1 Precipitation method

In this method, perovskite powders are commonly prepared for the analysis of their physical properties. To commence the process, $CH_3NH_3SnCl_3$ perovskite was synthesized by utilizing the solvents dimethylformamide (DMF) and dimethyl-sulfoxide (DMSO). A solution containing 0.1 M of methylammonium chloride was prepared separately using a mixture of DMF and DMSO in a 1:1 ratio. Additionally, a solution containing 0.1 M of tin chloride was prepared individually with the same solvents. Subsequently, the tin chloride solution was gradually introduced into the methylammonium chloride solution at a constant temperature of 60°C. The resultant product was filtered using polytetrafluoroethylene (PTFE) filter paper with a pore size of 0.2 μm. Finally, the obtained product was subjected to heating in a hot air oven at 100°C for 12 hours, followed by 120°C for 24 hours to remove the solvents from the perovskite [12, 13].

2.2 Ball milling method

This method is also used to prepare perovskite powders. The $CH_3NH_3SnBr_3$ was produced in polycrystalline structure via mechanochemical synthesis (ball milling) using precise quantities of SnBr2 and CH_3NH_3Br. The combined mass of reactants equaled 1.5 g, meticulously blended with 20 zirconia balls (5 mm in diameter) within a nitrogen-filled glove box. The reaction occurred within a sealed zirconia-lined jar, under a nitrogen atmosphere, using a Retsch PM100 mill, operating for 4 hours at 400 rpm [14].

2.3 Vapor deposition method

To carry out vapor deposition at a high vacuum pressure of 1.5×10^{-6} Torr for synthesizing $CH_3NH_3SnBr_3$, we placed $SnBr_2$ and CH_3NH_3Br powders in separate quartz crucibles heated by tungsten filaments. Pure thin films of $SnBr_2$ or MABr were deposited using a single source for reference samples. Evaporation rates were monitored using two quartz crystal microbalances from INFICON Co. for each source [15].

2.4 Dip coating method

In this synthesis method, CH_3NH_3Cl and $SnCl_2$ were dissolved individually in DMF and stirred until clear solutions formed. These precursor solutions were then combined to form the perovskite solution. The glass substrates underwent thorough cleaning using standard procedures: ultrasonic bath in acetone, isopropyl alcohol (IPA), and Deionized (DI) water for 20 minutes each, followed by UV-Ozone treatment. No specialized instruments were required for the dip coating process, which involved immersing the pre-cleaned glass substrate into the perovskite solution and subsequent annealing [16].

2.5 Spin coating

For the synthesis of $CH_3NH_3SnI_3$, the methylammonium iodide (CH_3NH_3I) and SnI_2 were dissolved in dimethyl sulfoxide (DMSO) at a molar ratio of 1:1: The concentration of $CH_3NH_3SnI_3$ was 39.9 wt%. The combined solution was applied onto the substrate using spin coating at 3000 rpm for 60 s. Toluene was dripped onto the substrate during the spin coating process and then dried on a hot plate at 65°C for 1 h [17–19].

2.6 Top-seeded solution method

In the synthesis process, for the synthesis of $CH_3NH_3SnI_3$, either formamidinium acetate or newly prepared CH_3NH_3I was combined with SnO in a solution containing HI-H_3PO_2 at 75°C in an air environment. Seed crystals were acquired through spontaneous crystallization. The conditions for crystal growth were enhanced by controlling the growth temperature and selecting high-quality seed crystals. Phase purity was confirmed with no detection of impurities [20].

3. Methylammonium-tin halide perovskites

This section reviews the literature related to methylammonium-tin halides ($CH_3NH_3SnX_3$; X = I, Br, Cl).

3.1 Methylammonium-tin-iodide-based perovskites

In 2011, the first study investigated the structural and electrical properties of $CH_3NH_3SnI_3$ perovskite through density-functional theory (DFT) calculations. First-principles calculations reveal the temperature of the single-crystal electrical conductivity, which shows metallic behavior down to low temperatures. Thermoelectric power also indicates metallic behavior, suggesting spontaneous hole-doping during crystallization and enhancing conductivity [21]. In 2014, the literature review showed that $CH_3NH_3PbI_3$/$MAPb(I_{1-x}Cl_x)_3$ perovskites have been used in photovoltaic technologies. Relativistic effects are shown in the correct modeling of electronic, optical, and transport properties by a proposed effective GW approximation (Green's function G and the screened Coulomb interaction W) that incorporates spin-orbit coupling. Relativistic effects stabilize $CH_3NH_3PbI_3$ against oxidation but increase the band gap compared to $CH_3NH_3SnI_3$, impacting solar cell performance [22]. In 2015, previous research showed highly uniform and pinhole-free $CH_3NH_3SnI_3$ perovskite films are achieved by utilizing a transitional phase of SnI_2·3DMSO solution. The solar cells based on mesoporous TiO_2 layer exhibit unprecedented photocurrent up to 21 mA cm^{-2}. High carrier densities are revealed by charge extraction and transient photovoltaic degradation studies in the $CH_3NH_3SnI_3$ perovskite device. This research advances lead-free perovskite solar cell development through improved morphology control [23].

In 2016, previous studies showed high-pressure treatments on lead-free $CH_3NH_3SnI_3$ perovskite, revealing improved structural stability, electrical conductivity, and photo responsiveness. In situ synchrotron XRD and Raman spectroscopy show phase transitions and amorphization with increased conductivity and visible-light responsiveness under pressure. A transition from tetragonal to orthorhombic occurred at 0.7 GPa. Interestingly, no amorphization was observed during the re-compression phase. These findings suggest the potential for high-performance photovoltaic materials developed under ambient pressure, inspiring further research in optimizing hybrid perovskite films [9]. However, the study addresses challenges in Sn-based perovskite solar cells, utilizing a low-temperature vapor-assisted solution process (LT-VASP) for $CH_3NH_3SnI_3$ thin-film fabrication. Optimization of SnI_2 substrate temperature yields high-quality films, enabling solar cell fabrication with improved reproducibility and efficiency (1.9%). LT-VASP films exhibit lower doping levels, reducing film conductivity and overcoming short-circuit behavior seen in conventional methods, thus advancing Sn-perovskite photovoltaic research [10]. On the other hand, the study utilizes hybrid functional method to analyze electronic properties and carrier mobility of $CH_3NH_3SnI_3$ perovskite as a solar cell absorber. Results show anisotropic electron and hole mobilities with a large magnitude of 1.4×10^4 cm^2 V^{-1} s^{-1}, suggesting $CH_3NH_3SnI_3$ as a p-type semiconductor. The relationship between effective mass anisotropy and electronic occupation offers guidance for experimental applications in electronics and optoelectronics [24]. In contrast, the study reports obtaining bulk single crystals of lead-free $CH_3NH_3SnI_3$ and $CH(NH_2)_2SnI_3$ perovskites via the top-seeded solution growth method. Detailed analysis includes structural refinement, band gap, thermal properties, and X-ray photoelectron spectroscopy (XPS) measurements. $CH_3NH_3SnI_3$ single crystals are p-type semiconductors, while $FASnI_3$ single crystals are n-type semiconductors. The trap densities in these crystals are in the order of 10^{11} cm^3. The stability was measured, the stability of $FASnI_3$ is higher than that of $CH_3NH_3SnI_3$. These findings suggest potential applications of both perovskites in photovoltaics, offering a non-toxic alternative to lead-based materials [20].

However, lead-free perovskite solar cells (PSCs) using $CH_3NH_3SnI_3$ as an alternative to lead-based perovskites show promise due to their narrower band gap and wider absorption spectrum. The study investigates parameters influencing PSC performance through theoretical analysis and device simulation, suggesting that optimizing doping concentration (1.3×10^{16} cm^{-3}) and the defect density (1×10^{15} cm^{-3}) can enhance efficiency, achieving encouraging results of Jsc of 31.59 mA/ cm^2, Voc of 0.92 V, FF of 79.99%, and power conversion efficiency (PCE) of 23.36%. Further research focuses on improving Sn^{2+} stability and reducing defect density for future advancements in $CH_3NH_3SnI_3$ PSCs [25]. Nevertheless, the study investigates $CH_3NH_3SnI_3$ in lead-free perovskite solar cells, noting lower efficiencies compared to lead-based perovskites. Photoluminescence spectra and time-resolved photoluminescence (PL) reveal improved carrier lifetime and reduced photoluminescence (PL) linewidth below 110 K, indicating hindered performance related to the orthorhombic to tetragonal phase transition. Targeted structural changes could mitigate defects and enhance charge extraction. Additionally, an above-gap emission feature suggests the potential for excess energy harvesting [26]. In addition, the study investigates photoexcited dynamics in lead-free $CH_3NH_3SnI_3$ perovskite films, revealing dominant free carrier recombination similar to lead analog $CH_3NH_3PbI_3$. Hot carrier relaxation occurs in 0.5 ps. Fluorescence quenching yields long electron (279 ± 88 nm) and hole (193 ± 46 nm) diffusion lengths. SnF_2 doping enhances fluorescence lifetime and diffusion lengths, suggesting $CH_3NH_3SnI_3$ potential as a non-toxic lead-free alternative for solar cells [27]. The study discusses the characteristics of organic-inorganic perovskite solar cells with a layer structure of FTO/d-TiO_2/mp-TiO_2:$CH_3NH_3SnI_3$/ PTAA/Au. The device exhibited a significantly small Voc of 0.23 V, which is far below the optical band gap Eg = 1.23 eV obtained from the external quantum efficiency (EQE) spectra. The photon energy loss is evaluated to be as large as E_{loss} = 1.0 eV, which is almost two times larger than that reported for $CH_3NH_3PbI_3$ perovskite solar cells. The study investigates the source of this loss in $CH_3NH_3SnI_3$-based devices, finding that it primarily stems from surface recombination at the interface rather than bulk recombination within the perovskite material [17].

In 2017, previous studies emphasized the study uncovers a notable photoresponse in $CH_3NH_3PbI_3$/ $CH_3NH_3SnI_3$ heterojunctions, displaying unconventional behavior distinct from typical photodiodes. The response increases steadily over a wide range of incident flux, with a reversal in polarity under illumination. This suggests an incident light-induced insulator-to-metal transition in the $CH_3NH_3PbI_3$ side, similar to models used for the lead homolog, promising success in photovoltaic applications [28]. Moreover, the study investigates the optical responses and photocarrier dynamics of long-term stable $CH_3NH_3SnI_3$ thin films and solar cells. Films prepared with SnF_2 exhibit a bandgap of 1.25 eV, with improved photocarrier lifetime and device performance compared to SnF_2-free films, attributed to reduced non-radiative recombination [29]. In 2018, the study showed that the instability of $CH_3NH_3SnI_3$ perovskite is investigated through photoelectron spectroscopy and theoretical calculations. Results reveal changes in surface chemical states, broken Sn-I bonds, and depletion of $CH_3NH_3^+$ cations, suggesting that surface chemical instability is a key factor in the material's instability [30]. In 2019, the study demonstrated a hybrid methylamine iodide and Pb-Sn binary perovskite for photodetection, extending the photoresponse onset wavelength to 1000 nm with 30 mol % Sn content. The photodetectors exhibit a photoresponsivity of 0.39 A W^{-1}, a specific detectivity of 7×10^{12} Jones, a fast photoresponse with rise and decay time constants, and an external quantum efficiency greater than 50% in the wavelength range of 350–900 nm, with a maximum value of about 80% at 550 nm [31]. Moreover, the study $CH_3NH_3SnI_3$ crystals grow up to 10 mm × 6 mm × 6 mm in aerobic conditions using cooling crystallization. X-ray diffraction (XRD) confirms cubic phase with lattice constants a = b = c = 6.2480 Å. HRTEM reveals good crystallinity. FT-IR spectrum and band gap estimation (1.33 eV) are provided. Thermal analysis and first-principles calculations support potential solar cell applications despite limited chemical stability [32].

In 2020, the study investigated $CH_3NH_3SnI_3$ perovskite's electronic and optical behavior via density-functional theory. Calculations using PW-LDA, GGA (PBE), and GW methods yield varying band gap values (0.77–1.38 eV). Sn and I atoms significantly influence electronic properties. The bandgap values obtained by GGA (1.02 eV) are bigger than that obtained by LDA (0.77 eV), and the best band gap value for the $CH_3NH_3SnI_3$ perovskite obtained by the GW approximation (1.38 eV), which is close to the experimental value. Optical analysis suggests incident photon energy in 1.5–3 eV energy range of $CH_3NH_3SnI_3$ suitability as a solar cell active layer [33]. In 2021, the study explored the potential of lead-free $CH_3NH_3SnI_3$ perovskite solar cells using ZnTe as a hole-transport layer. Through modeling, simulation, and fabrication, efficiencies of 22.96% (simulation) and 8.41% (fabrication) are achieved, showcasing the viability of this approach for cost-effective, environmentally friendly solar cell production [18]. Furthermore, the study simulates a photoexcited $CH_3NH_3SnI_3$-based solar cell device using a capacitance simulator. Efforts focus on absorber layer thickness, acceptor concentration, defect density, and metal contact work function impact on photovoltaic properties. Further, optimizing the defect density (1014 cm^{-3}) of the perovskite absorber layer, encouraging results of the Jsc of 40.14 mA/cm^2, Voc of 0.93 V, FF of 75.78%, and PCE of 28.39% was achieved. An anode material with a high work function is deemed necessary for improved device performance [34].

In 2022, the previous studies were limited to $CH_3NH_3SnI_3$ perovskites in thermoelectric applications. Applying high pressure with shear strain enhances structural stability, electrical conductivity, and Seebeck coefficient in $CH_3NH_3SnI_3$ tube structures. First-principles calculations confirm changes in properties. This presents a promising strategy to improve $CH_3NH_3SnI_3$ thermoelectric properties [35]. Additionally, the study explores enhancing perovskite solar cell performance by reducing defects. SCAPS-1D simulations evaluate efficiency limits and propose optimal architectures. Selection of hole-transport material and electrode is crucial, with a particular focus on minimizing bulk and interface defects in $CH_3NH_3SnI_3$ layers. Finally, the optimal device architecture $TiO_2/CH_3NH_3SnI_3/Cu_2O$ is proposed for the given absorber layer using the readily available Cu_2O hole-transporting material with PCE = 27.95%, FF = 84.0 5%, Voc = 1.02 V and JSC = 32.60 mA/cm^2, providing optimal performance and enhanced resistance to defects [36]. In 2023, the study investigated $CH_3NH_3SnI_3$ perovskite solar cells with various electron transport layers (ETLs) and hole-transport layers (HTLs) using SCPAS 1D software. It identifies TiO_2 as the most suitable ETL and Cu_2O as the optimal HTL. ZnO is chosen as ETL due to its advantages over TiO_2. The interface defect at the ETL and HTL junctions has also been analyzed by varying its concentration in the range of 10^{11}–10^{16} cm^{-3} Analysis includes the impact of absorber thickness and defects, interface defect concentration, acceptor/donor density variations, ETL electron affinity, back contact metal work function, temperature, and series resistance on solar cell performance, aiming for higher efficiency and sustainable energy production [37]. Moreover, the study tackles the synthesis challenge of lead-free $CH_3NH_3SnI_3$ perovskite for sustainable energy devices, emphasizing preventing Sn^{2+} oxidation during synthesis. A modified nanoparticle synthesis in an inert environment produces crystalline $CH_3NH_3SnI_3$ perovskite. The structural parameters such as crystalline size (D), dislocation density (δ), strain (ε), and the lattice parameter were calculated from the XRD pattern and were found to be 62.90 nm, 2.52×10^{14} nm^{-2}, 6.12×10^{-3} lin^2/m^{-4}, and 6.23951 Å, respectively. The conductivity relaxation time, DC conductivity, and AC conductivity were calculated as 6.45×10^{-5} s, 2.01×10^{-6} S/m, and 6.98×10^{-4} S/m, respectively. Characterization confirms its properties, including electrical conductivity and performance in Schottky devices, showing promise for future applications [38]. Furthermore, the study enhances charge conduction in $CH_3NH_3SnI_3$ perovskite by incorporating single-walled carbon nanotubes (SWCNTs) to mitigate tin oxidation and trap states. The synthesized perovskite-SWCNT composites show improved device performance, with increased conductivity and reduced series resistance. Specifically, in the absence of light, SWCNTs increase the conductivity from 3.8×10^{-5} to 1.6×10^{-4} S/m and decrease the series resistance from 348.25 to 34.80 Ω. Under illumination, SWCNTs further increase conductivity to 4.0×10^{-4} S/m from 8.1×10^{-5} S/m and decrease series resistance to 19.20 Ω from 220.56 Ω. SWCNTs act as charge extraction channels and trap state fillers, promising high-performance optoelectronic devices [39].

3.2 Methylammonium-tin-bromide based perovskites

In 1988, the first study of $CsSnBr_3$ and $CH_3NH_3SnBr_3$ perovskite exhibited cubic structures at room temperature and successive phase transitions at lower temperatures, as observed through ^{81}Br NQR and differential thermal analysis (DTA). Structural distortions, interpreted via the three-center-four-electron bond model, were evidenced by ^{81}Br NQR and ^{119}Sn Moessbauer spectra. Dynamic properties of the cation were assessed through broad-line ^{1}HNMR [40]. After that, in 1991, the heat capacity study of $CH_3NH_3SnBr_3$ revealed four anomalies at temperatures: 46.0 K, 188.2 K, 213.0 K, and 229.4 K, corresponding to higher-order (H) and first-order (F) transitions. Transition entropies suggest displacive and order-disorder types, consistent with methylammonium ions orientational disorder. Electric properties shift from insulating to semiconducting at 229.4 K, with a conductivity increase of 10^4. Thermal activation energy for conductivity is 29.1 kJ/mol. Dielectric dispersion below 150 K exhibits a complex relaxation behavior [41]. In 2006, it was reported in literature that $CH_3NH_3SnBr_3$ hybrid films, deposited by single-source thermal ablation, exhibited cubic perovskite crystallization and c-axis orientation on various substrates. Resistivity measurements indicated semiconductor behavior, with an activation energy of 0.30 eV. Hole-type charge carriers with a room temperature mobility of ~10^{-5} cm^2 V^{-1} s^{-1} were observed, increasing exponentially with gate voltage at higher temperatures [42]. In 2010, the synchrotron powder diffraction data of $CH_3NH_3SnBr_3$ revealed a phase transition between 230 and 188 K, crystallizing in $Pmc2_1$ with strong ferroelectric distortions. The Sn $5s^2$ lone pair drives octahedral tilting and a ferroelectric mode. The lower-temperature phase is likely triclinic, with density-functional theory suggesting the amine groups' directional bonding to the halide cage influences the inorganic sublattice, particularly the Sn lone pair direction, at little energy cost [43].

In 2016, lead-free $CH_3NH_3SnBr_3$ perovskite films were deposited via co-evaporation and sequential evaporation methods. Structural and electronic properties were characterized, revealing crystalline structure and an optical bandgap of ~2.2 eV. Solar cells employing such perovskite film in a planar structure were fabricated with various hole-transport layers (HTLs) (spiro-OMeTAD, C60, and P3HT). Solar cells using sequential deposition achieved higher efficiencies (up to 1.12%) obtained using P3HT as the HTL [44]. In 2018, the method introduced on lead-free $CH_3NH_3SnBr_{3-x}I_x$ (x = 0–3) perovskite quantum dots (QDs) with tunable optical bandgaps was developed for mesoscopic solar cells. These QDs, when applied as light harvesters, exhibit reasonable power conversion efficiency (8.79%) and rapid start-up behaviors in conjunction with N719 co-sensitized mesoscopic solar cells [45]. In 2019, the investigation showed that lead-free halide perovskite candidates, including $CH_3NH_3SnBr_3$, $CH(NH_2)_2SnBr_3$, $CH_3NH_3SnI_3$, and $CH(NH_2)_2SnI_3$, were characterized using DTA, XRD, NQR, and NMR techniques. A novel synthetic method utilizing pure water was proposed. $CH(NH_2)_2SnBr_3$ exhibited a unique lattice constant, and SnF_2 aided in detecting ^{119}Sn NMR signals effectively [46]. In 2021, the study showed that lead-free $CH_3NH_3SnBr_3$ perovskite, a promising alternative to toxic lead-based counterparts, shows a band gap of 1.9 eV, suitable for photovoltaic applications across visible to ultraviolet wavelengths. Prepared via ball milling, its structural transitions down to 120 K were elucidated by synchrotron X-ray diffraction. Distinct changes in the orthorhombic framework, attributed to $SnBr_6$ octahedral tilting and H-bond interactions with $CH_3NH^+{}_3$, were correlated with calorimetric events. With a Seebeck coefficient of around 2000 mV K^{-1} near room temperature, it outperforms other halide perovskites, promising advancements in solar energy conversion [14].

3.3 Methylammonium-tin-chloride-based perovskites

In 2008, the first study synthesized well-oriented thin films of methylammonium trihalogenostannates ($CH_3NH_3SnX_3$) using single-source thermal ablation. Structural and optical analyses, combined with ab initio calculations, elucidate electronic properties. Results suggest the band gap depends on the inorganic cage's ionicity and size rather than the organic cation [47]. Moreover, another study investigates the structural and electronic properties of inorganic and hybrid organic–inorganic perovskites based on tin halides using first principles. By contrasting $CsSnCl_3$ and $CsSnI_3$ with their hybrid counterparts $CH_3NH_3SnCl_3$, $CH_3NH_3SnI_3$, and $NH_2CH{=}NH_2SnI_3$, insights into the effects of embedded cations on stability and electronic properties are revealed. The choice of organic molecule, particularly its size, allows for tunability of electronic properties such as band gap [48].

In 2011, It was reported in the literature that impedance spectroscopy of thermally ablated $CH_3NH_3SnCl_3$ films reveals enhanced complex conductivity with rising humidity. Coplanar two-electrode devices exhibit resistive influence from ambient humidity. Dynamic responses show significant current variation across humidity levels, suggesting surface-bound water molecules drive impedance changes, likely through proton conduction [49]. In 2014, using the HSE06 hybrid functional, the study examined $CH_3NH_3SnX_3$ (X = Cl, Br, I) perovskites as solar cell absorbers. Computed direct band gaps at the Γ point range from 1.67 to 3.0 eV. Carrier effective masses and band gaps decrease from Cl to I. Photon absorption efficiency slightly improves under isotropic compressive stress, reducing the fundamental band gap [50]. In 2016, the study focused on non-toxic perovskite halide $CH_3NH_3SnCl_3$, calculating crystal structures, optical properties, and carrier mobility using density-functional theory with van der Waals correlation. Results reveal a high optical absorption coefficient (10^5 cm^{-1}) and excellent solar light harvesting capability. Triclinic phase exhibits significant electron mobility (1700 $cm^2\,V^{-1}\,s^{-1}$), attributed to small effective mass and ferroelectricity from distorted $SnCl_6$ octahedra [51].

In 2017, the literature review shows that pressure-induced structural evolution, optical, and electronic transitions in non-toxic organometal halide perovskite-based $CH_3NH_3SnCl_3$. The high-pressure behavior of MA cations indicates that the interaction between MA cations and $[SnCl_6]^4$ octahedra is strengthened. The pressure-induced electrical resistance evolution of $MASnCl_3$ coincides with the structural changes. The pressure-induced perovskites show potential for efficient solar cell power conversion without toxic lead components [52]. Moreover, the study based on methylammonium ($CH_3NH^-{}_3$) mixed $CH_3NH_3SnIBr_{2-x}Cl_x$ was synthesized and characterized for lead-free, carbon-based mesoscopic solar cells. Varying $SnCl_2$/$SnBr_2$ ratios resulted in cocrystallized perovskites with three halides for $x \leq 0.5$, while phase separation occurred for $x \geq 1$. A device with $CH_3NH_3SnIBr_{1.8}Cl_{0.2}$ ($SnCl_2$ = 10%) exhibited the best photovoltaic performance, J_{SC}/mA cm^{-2} = 14.0, V_{oc}/mV =380, FF = 0.573 and PCE/% = 3.1, with reproducibility and long-term stability. Electrochemical impedance spectra showed superior charge recombination and dielectric relaxation properties for this cell. Transient PL decays indicated that the $CH_3NH_3SnIBr_{1.8}Cl_{0.2}$ film had the greatest lifetime, contributing to its superior performance [53]. In 2018, the paper introduced a novel, low-cost, and non-toxic perovskite sensitizer material, $CH_3NH_3SnCl_3$, for photovoltaic applications. Synthesized via direct deposition of equimolar CH_3NH_3Cl and $SnCl_2$ in DMF solution, its perovskite structure is confirmed by XRD. UV-vis absorption spectroscopy reveals optical bandgap of perovskite was estimated to be 2.5 eV for powder perovskite and 2.1 eV for crystal perovskite. The fabricated perovskite solar cell with PEO-based solid polymer electrolyte shows reproducible PCE of 0.17% in case of powder perovskite, and finally, we got 0.55% in case of crystal perovskite of $CH_3NH_3SnCl_3$ at 100 mW/cm^2 (1 sun condition) processed in the ambient air surroundings. Comparative analysis of crystal and powder forms explores structural, optical, and electrical properties in depth, presenting its potential for perovskite solar cells [19]. On the other hand, the study explores perovskite sensitizer materials-based solar cells (PSSC), focusing on fabrication and characterization of $CH_3NH_3SnCl_3$ perovskite. PSSC is assembled with polyethylene oxide solid polymer electrolyte. Fabricated solar cell exhibits Jsc of 1.8 mA/cm^2, Voc of 0.48 V, fill factor of 0.68, and efficiency of 0.60% under 1 sun condition [54]. In addition, the study proposes replacing Pb with tin (Sn) to fabricate $CH_3NH_3SnCl_3$ cells in a chlorine (Cl)-rich environment, enhancing film morphology and reducing oxidation. Characterization techniques including X-ray diffraction and SEM examine annealing temperature effects. The photovoltaic performance is evaluated under solar simulator illumination, the best performance recorded at 110°C, Voc/mV = 515, Jsc/mA cm^{-2} = 9.60, P_{max} = 2.76, FF = 0.55, PCE = 2.30% [55].

In 2019, methylammonium-tin-chloride perovskite thin films were produced via solution processing methods (spin coating, dip coating). Characterization via powder XRD confirmed cubic phase, while optical band gaps (3.69 eV) were estimated using Taucs plot. Photoluminescence studies showed higher intensity in vacuum-annealed films, indicating fewer defects. Comparing vacuum annealing at 150°C and 200°C, the former exhibited fewer defects, suggesting optimal annealing conditions for practical device applications [16]. However, the study compares the photovoltaic performance of $Ba_{0.94}La_{0.06}SnO_3$ and TiO_2 nanoparticles synthesized through different methods alongside NiO and $CH_3NH_3SnCl_3$ nanoparticles. XRD patterns confirm cubic, tetragonal, and cubic phases with average sizes ~25 nm. $CH_3NH_3SnCl_3$ shows a monoclinic phase with an average size of ~5 nm. Band gaps are 3.6 eV for $Ba_{0.94}La_{0.06}SnO_3$, 3.4 eV for TiO_2, 3.5 eV for NiO, and 2.8 eV for $CH_3NH_3SnCl_3$. Thermally deposited silver contacts serve as counter electrodes, with device performance evaluated under ~100 mW cm^2 illumination for power conversion efficiency [56].

In 2020, prior research suggested that solvent engineering is crucial for optimizing organic-inorganic halide perovskites (OIHPs) in optoelectronic applications. $CH_3NH_3SnCl_3$ perovskites were synthesized using DMF:DMSO and DMF:ACN solvent mixtures, showing varied structural, morphological, and optical properties. Field emission scanning electron microscopy (FESEM) imaging, XRD, TGA, UV-Vis, PL spectroscopy, and DFT calculations elucidated solvent-dependent effects, with DMF:DMSO exhibiting superior properties [12]. In contrast to this, the study investigates the properties of 2D halide perovskites ($CH_3NH_3ZX_3$) using density-functional theory. It reveals that an increasing halogen atomic radius from Cl to I leads to a larger lattice constant and a reduced bandgap, with $CH_3NH_3PbX_3$ having a higher bandgap than $CH_3NH_3SnX_3$. Optical analysis shows $CH_3NH_3PbX_3$ superior static dielectric constant, and both materials exhibit optical activities in visible and high ultraviolet regions with a high absorption coefficient. The 2D perovskites demonstrate higher carrier mobility than bulk systems, especially $CH_3NH_3PbCl_3$, $CH_3NH_3PbBr_3$, and $CH_3NH_3SnCl_3$. Lead-free $CH_3NH_3SnBr_3$ exhibits a higher efficiency (28.62%) than $CH_3NH_3PbI_3$ (28.21%), suggesting potential applications in optoelectronics and solar cells [57]. On the other hand, organic-inorganic perovskite solar cells exhibit high promise due to their increasing efficiency, material abundance, low cost, and simple synthesis. $CH_3NH_3SnCl_3$ and $CH_3NH_3ISnCl_2$ are synthesized via chemical methods for their simplicity. Structural and morphological characterization via XRD and SEM reveals cubic and tetragonal phases for $CH_3NH_3SnCl_3$ and $CH_3NH_3ISnCl_2$, respectively, with average grain sizes around 40.9 nm and 40.62 nm. SEM images show irregular crystal growth with average particle sizes of approximately 334.68 μm for $CH_3NH_3SnCl_3$ and 464.88 μm for $CH_3NH_3ISnCl_2$ [58].

Compound	Method	Device architecture	Structural phase	Bandgap (eV)	Efficiency (%)	Reference
$CH_3NH_3SnI_3$	Vapor-assisted solution process	FTO/c-TiO_2/mp-TiO_2/$CH_3NH_3SnI_3$/PTAA/TPFB /Au	Pseudocubic	1.30	1.86	[10]
	Spin coating	FTO/d-TiO_2/mp-TiO_2/$CH_3NH_3SnI_3$/PTAA/Au	Pseudocubic	1.23	2.33	[17]
	Spin coating	FTO/TiO_2/$CH_3NH_3SnI_3$/ZnTe/FTO	Orthohombic	1.23	8.41	[18]
	Solvent mediated crystallization	FTO/c-TiO_2/mp-TiO_2/$CH_3NH_3SnI_3$/PTAA/Au	Tetragonal	1.55	5.79	[23]
$CH_3NH_3SnBr_3$	Vapor deposition	FTO/c-TiO_2/mp-TiO_2/$CH_3NH_3SnI_3$/PTAA/Li-TFSI/ P3HT/Au	Cubic	2.20	1.12	[14]
$CH_3NH_3SnBr_{1.8}Cl_{0.2}$	Drop casting	FTO/c-TiO_2/mp-TiO_2/$CH_3NH_3SnBr_{1.8}Cl_{0.2}$/$Al_2O_3$/C	Tetragonal	1.97	3.10	[53]
$CH_3NH_3SnCl_3$	Spin coating	FTO/b-TiO_2/n-TiO_2/ $CH_3NH_3SnCl_3$/PEO/Pt	Monoclinic	2.10	0.50	[19]
	Spin coating	FTO/b-TiO_2/n-TiO_2/ $CH_3NH_3SnCl_3$/PEO/Pt	Triclinic	2.27	0.60	[54]
	Spin coating	FTO/c-TiO_2/$CH_3NH_3SnCl_3$/CuSCN/Ag	Triclinic	2.30	1.61	[55]
	Spin coating	FTO/$Ba_{0.94}La_{0.06}SnO_3$/$CH_3NH_3SnCl_3$/NiO/Ag	Monoclinic	2.80	1.90	[56]

Table 1.
Comparative study of structural phase, optical bandgap, and device efficiency of $CH_3NH_3SnX_3$; X = I, Br, Cl via experimental approach.

In 2023, the literature review showed that methylammonium-tin halide ($(CH_3NH_3)_2SnX_6$) double perovskite thin films offer stability and are integrated into self-charging power units (SPUs). ($(CH_3NH_3)_2SnCl_6$-based piezoelectric nanogenerators (PENGs) achieve high power density, while ($(CH_3NH_3)_2SnCl_6$-based lithium metal batteries (LMBs) demonstrate high specific capacity and stability with Al_2O_3 coating. Collectively, these components drive small-scale electronics, demonstrating the viability of metal halide perovskites for efficient and sustainable energy systems, and informing future principles in device design [59]. In contrast to this, the literature discusses the synthesis of organic-inorganic hybrid perovskites, including $(CH_3NH_3)_2CuCl_4$, $CH_3NH_3PbCl_3$, and $CH_3NH_3SnCl_3$, utilizing an ultrasonic radiation-assisted method. X-ray diffraction analysis confirms the presence of crystalline phases in these perovskites. The UV-Vis diffuse reflectance spectroscopy reveals a band gap in $CH_3NH_3PbCl_3$ (ABX_3) varied from 2.03 to 2.0 eV and 2.03 to 2.3 eV, respectively, when Cu and Sn are replaced for the B sites. Current-voltage characteristics reveal increased photocurrent under UV light compared to darkness, indicating improved performance in the visible absorption region, consistent with previous research. The study emphasizes the photo-induced mechanisms responsible for the remarkable charge separation efficiency of these materials, highlighting their significance in advancing photovoltaic research [60]. In 2024, previous studies have emphasized lead-free methylammonium-tin chlorides ($CH_3NH_3SnCl_3$, $CH_3NH_3Sn_2Cl_5$, and $(CH_3NH_3)_4SnCl_6$) were synthesized and characterized for structural, optical, electrochemical, and thermal properties. XRD, FE-SEM, FT-IR, UV-Vis DRS, PL, TGA, and CV analyses revealed their crystal structures, morphology, semiconducting behavior, emission centers, thermal kinetics, and supercapacitor performance, with $CH_3NH_3Sn_2Cl_5$ showing superior results [13]. The comparative study of structural phase, optical bandgap and device efficiency of $CH_3NH_3SnX_3$; X = I, Br, Cl via experimental approach is tabulated in **Table 1**.

4. Conclusion

In summary, this chapter reveals the literature regarding the progress and development of methylammonium-tin halides (I, Br, Cl) perovskites. The perovskites were prepared via precipitation methods, ball milling, vapor deposition, dip coating, spin coating and top-seeded solution method. Various synthesis strategies were proposed, such as pressure, solvent, composition variations, and the incorporation of SnF_2, SnCl, and SWCNT. The methods and strategies employed impacted the structural, morphological, optical, electrical, and thermal properties of the perovskites. Additionally, perovskites were fabricated as light harvesters in solar photovoltaic (solar PV), with reported efficiencies of 28.30% (simulation) and 8.41% (fabrication), respectively. The literature suggests that the replacement of lead ions with tin is suitable, as tin is a low-toxicity metal. However, the instability problem of tin-based devices still remains. Various intrinsic and extrinsic factors affect the stability of the devices, such as chemical structure, oxidation, moisture, heat, trap states, and defects. This report shows that oxidation can be reduced by the addition of SnF_2, Cl^-, and SWCNT. The selection of hole-transport materials and electrodes reduces interface defects, thereby increasing device stability and lifetime. However, the methylammonium tin halides (I, Br, Cl) perovskites are potential for solar PV and optoelectronic devices.

Author details

Nagaraj Gopinathan and Syed Abdul Gafoor Sathik Basha*
Department of Physics, B.S. Abdur Rahman Crescent Institute of Science and Technology, Chennai, India

*Address all correspondence to: sathiknoushin@gmail.com

References

[1] Tian J, Xue Q, Yao Q, Li N, Brabec CJ, Yip H. Inorganic halide perovskite solar cells: Progress and challenges. Advanced Energy Materials. 2020;**10**(23):2000183 Available from: https://onlinelibrary.wiley.com/doi/10.1002/aenm.202000183

[2] Zhao S, Mo Q, Wang B, Cai W, Li R, Zang Z. Inorganic halide perovskites for lighting and visible light communication. Photonics Research. 2022;**10**(4):1039. Available from: https://opg.optica.org/abstract.cfm?URI=prj-10-4-1039

[3] Wang H, Li S, Liu X, Shi Z, Fang X, He J. Low-dimensional metal halide perovskite photodetectors. Advanced Materials. 2021;**33**(7):2003309 Available from: https://onlinelibrary.wiley.com/doi/10.1002/adma.202003309

[4] Wu Z, Yang J, Sun X, Wu Y, Wang L, Meng G, et al. An excellent impedance-type humidity sensor based on halide perovskite CsPbBr3 nanoparticles for human respiration monitoring. Sensors and Actuators B: Chemical. 2021;**337**:129772. Available from: https://linkinghub.elsevier.com/retrieve/pii/S0925400521003415

[5] Peedikakkandy L, Bhargava P. Composition dependent optical, structural and photoluminescence characteristics of cesium tin halide perovskites. RSC Advances. 2016;**6**(24):19857-19860. Available from: http://xlink.rsc.org/?DOI=C5RA22317B

[6] Chergui M. Localized holes and delocalized electrons in photoexcited inorganic perovskites: Watching each atomic actor by picosecond X-ray absorption spectroscopy localized holes and delocalized electrons in photoexcited inorganic perovskites: Watching each atomic actor by picosecond X-ray absorption spectroscopy. Structural Dynamics. 2017;**4**:044002

[7] Naresh V, Singh S, Soh H, Lee J, Lee N. Dual-phase CsPbBr3–CsPb2Br5 perovskite scintillator for sensitive X-ray detection and imaging. Materials Today Nano. 2023;**23**:100364. Available from: https://linkinghub.elsevier.com/retrieve/pii/S2588842023000639

[8] Li J, Duan J, Yang X, Duan Y, Yang P, Tang Q. Review on recent progress of lead-free halide perovskites in optoelectronic applications. Nano Energy. 2021;**80**:105526. Available from: https://linkinghub.elsevier.com/retrieve/pii/S2211285520311009

[9] Lü X, Wang Y, Stoumpos CC, Hu Q, Guo X, Chen H, et al. Enhanced structural stability and photo responsiveness of $CH_3NH_3SnI_3$ perovskite via pressure-induced amorphization and recrystallization. Advanced Materials. 2016;**28**(39):8663-8668. Available from: https://onlinelibrary.wiley.com/doi/10.1002/adma.201600771

[10] Yokoyama T, Cao DH, Stoumpos CC, Song TB, Sato Y, Aramaki S, et al. Overcoming short-circuit in lead-free $CH_3NH_3SnI_3$ perovskite solar cells via kinetically controlled gas–solid reaction film fabrication process. Journal of Physical Chemistry Letters. 2016;**7**(5):776-782. Available from: https://pubs.acs.org/doi/10.1021/acs.jpclett.6b00118

[11] Wu T, Liu X, Luo X, Lin X, Cui D, Wang Y, et al. Lead-free tin perovskite solar cells. Joule. 2021;**5**(4):863-886. Available from: https://linkinghub.elsevier.com/retrieve/pii/S2542435121000969

[12] Gopinathan N, Basha SS, Banu IBS, Mamat MH, Sirajudeen MMS. Solvents driven structural, morphological, optical and dielectric properties of lead free perovskite $CH_3NH_3SnCl_3$ for optoelectronic applications: Experimental and DFT study. Materials Research Express. 2020;**6**(12):125921. Available from: https://iopscience.iop.org/article/10.1088/2053-1591/ab6a47

[13] Gopinathan N, Basha SS, Vasimalai N, Mundari NAA, Shajahan A, Parveen JS, et al. Composition-driven structural, optical, thermal and electrochemical properties of hybrid perovskite-structured methylammonium-tin-chloride. Journal of Electronic Materials. 2024;**53**(1):94-105. Available from: https://link.springer.com/10.1007/s11664-023-10777-0

[14] López CA, Abia C, Gainza J, Kayser P, Nemes NN, Dura OJ, et al. Structural evolution, optical gap and thermoelectric properties of $CH_3NH_3SnBr_3$ hybrid perovskite, prepared by mechanochemistry. Materials Advances. 2021;**2**(11):3620-3628. Available from: http://xlink.rsc.org/?DOI=D1MA00196E

[15] Jung MC, Raga SR, Qi Y. Properties and solar cell applications of Pb-free perovskite films formed by vapor deposition. RSC Advances. 2016;**6**(4):2819-2825. Available from: http://xlink.rsc.org/?DOI=C5RA21291J

[16] Kodibailu A, Satyanarayan MN. Study of the effect of annealing parameters on the perovskite material (CH3NH3SnCl3) using PL spectra. In Hisar, Haryana, India; 2019. p. 030587. Available from: http://aip.scitation.org/doi/abs/10.1063/1.5113426

[17] Do Kim H, Miyamoto Y, Kubota H, Yamanari T, Ohkita H. Open-circuit voltage loss in $CH_3NH_3SnI_3$ perovskite solar cells. Chemistry Letters. 2017;**46**(2):253-256. Available from: https://academic.oup.com/chemlett/article/46/2/253-256/7337755

[18] Kumari K, Jana A, Dey A, Chakrabarti T, Sarkar SK. Lead free CH3NH3SnI3 based perovskite solar cell using ZnTe nano flowers as hole transport layer. Optical Materials. 2021;**111**:110574. Available from: https://linkinghub.elsevier.com/retrieve/pii/S0925346720309149

[19] Rahul SPK, Singh R, Singh V, Bhattacharya B, Khan ZH. New class of lead free perovskite material for low-cost solar cell application. Materials Research Bulletin. 2018;**97**:572-577. Available from: https://linkinghub.elsevier.com/retrieve/pii/S0025540817325382

[20] Dang Y, Zhou Y, Liu X, Ju D, Xia S, Xia H, et al. Formation of hybrid perovskite tin iodide single crystals by top-seeded solution growth. Angewandte Chemie, International Edition. 2016;**55**(10):3447-3450. Available from: https://onlinelibrary.wiley.com/doi/10.1002/anie.201511792

[21] Takahashi Y, Obara R, Lin ZZ, Takahashi Y, Naito T, Inabe T, et al. Charge-transport in tin-iodide perovskite CH3NH3SnI3: Origin of high conductivity. Dalton Transactions. 2011;**40**(20):5563. Available from: http://xlink.rsc.org/?DOI=c0dt01601b

[22] Umari P, Mosconi E, De Angelis F. Relativistic GW calculations on CH3NH3PbI3 and CH3NH3SnI3 perovskites for solar cell applications. Scientific Reports. 2014;**4**(1):4467. Available from: https://www.nature.com/articles/srep04467

[23] Hao F, Stoumpos CC, Guo P, Zhou N, Marks TJ, Chang RPH, et al.

Solvent-mediated crystallization of $CH_3NH_3SnI_3$ films for heterojunction depleted perovskite solar cells. Journal of the American Chemical Society. 2015;**137**(35):11445-11452. Available from: http://pubs.acs.org/doi/10.1021/jacs.5b06658

[24] Wu LJ, Zhao YQ, Chen CW, Wang LZ, Liu B, Cai MQ. First-principles hybrid functional study of the electronic structure and charge carrier mobility in perovskite $CH_3NH_3SnI_3$. Chinese Physics B. 2016;**25**(10):107202. Available from: https://iopscience.iop.org/article/10.1088/1674-1056/25/10/107202

[25] Du HJ, Wang WC, Zhu JZ. Device simulation of lead-free $CH_3NH_3SnI_3$ perovskite solar cells with high efficiency. Chinese Physics B. 2016;**25**(10):108802. Available from: https://iopscience.iop.org/article/10.1088/1674-1056/25/10/108802

[26] Parrott ES, Milot RL, Stergiopoulos T, Snaith HJ, Johnston MB, Herz LM. Effect of structural phase transition on charge-carrier lifetimes and defects in $CH_3NH_3SnI_3$ perovskite. Journal of Physical Chemistry Letters. 2016;**7**(7):1321-1326. Available from: https://pubs.acs.org/doi/10.1021/acs.jpclett.6b00322

[27] Ma L, Hao F, Stoumpos CC, Phelan BT, Wasielewski MR, Kanatzidis MG. Carrier diffusion lengths of over 500 nm in Lead-free perovskite $CH_3NH_3SnI_3$ films. Journal of the American Chemical Society. 2016;**138**(44):14750-14755. Available from: https://pubs.acs.org/doi/10.1021/jacs.6b09257

[28] Spina M, Mihály L, Holczer K, Náfrádi B, Pisoni A, Forró L, et al. Photodiode response in a $CH_3NH_3PbI_3$/$CH_3NH_3SnI_3$ heterojunction. ACS Applied Materials & Interfaces. 2017;**9**(11):10198-10202. Available from: https://pubs.acs.org/doi/10.1021/acsami.6b12392

[29] Handa T, Yamada T, Kubota H, Ise S, Miyamoto Y, Kanemitsu Y. Photocarrier recombination and injection dynamics in long-term stable Lead-free $CH_3NH_3SnI_3$ perovskite thin films and solar cells. Journal of Physical Chemistry C. 2017;**121**(30):16158-16165. Available from: https://pubs.acs.org/doi/10.1021/acs.jpcc.7b06199

[30] Lee YM, Park J, Yu BD, Hong S, Jung MC, Nakamura M. Surface instability of Sn-based hybrid perovskite thin film, $CH_3NH_3SnI_3$: The origin of its material instability. Journal of Physical Chemistry Letters. 2018;**9**(9):2293-2297. Available from: https://pubs.acs.org/doi/10.1021/acs.jpclett.8b00494

[31] Wang Y, Yang D, Ma D, Kim DH, Ahamad T, Alshehri SM, et al. Organic-inorganic hybrid Sn-based perovskite photodetectors with high external quantum efficiencies and wide spectral responses from 300 to 1000 nm. Science China Materials. 2019;**62**(6):790-796. Available from: http://link.springer.com/10.1007/s40843-018-9377-3

[32] Huang YQ, Su J, Li QF, Wang D, Xu LH, Bai Y. Structure, optical and electrical properties of $CH_3NH_3SnI_3$ single crystal. Physica B: Condensed Matter. 2019;**563**:107-112. Available from: https://linkinghub.elsevier.com/retrieve/pii/S092145261930211X

[33] Sabetvand R, Ghazi ME, Izadifard M. DFT study of electronic and optical properties of $CH_3NH_3SnI_3$ perovskite. Energy Sources, Part A: Recovery, Utilization, and Environmental Effects. 2020:1-13. Available from: https://www.tandfonline.com/doi/full/10.1080/15567036.2020.1805047

[34] Patel PK. Device simulation of highly efficient eco-friendly $CH_3NH_3SnI_3$ perovskite solar cell. Scientific Reports. 2021;**11**(1):3082. Available from: https://www.nature.com/articles/s41598-021-82817-w

[35] Wang Q, Tang Y, Horita Z, Iikubo S. Structural and thermoelectric properties of $CH_3NH_3SnI_3$ perovskites processed by applying high pressure with shear strain. Materials Research Letters. 2022;**10**(8):521-529. Available from: https://www.tandfonline.com/doi/full/10.1080/21663831.2022.2057821

[36] Omarova Z, Yerezhep D, Aldiyarov A, Tokmoldin N. In silico investigation of the impact of hole-transport layers on the performance of $CH_3NH_3SnI_3$ perovskite photovoltaic cells. Crystals. 2022;**12**(5):699. Available from: https://www.mdpi.com/2073-4352/12/5/699

[37] Tiwari C, Gupta GK, Mishra V. Computational analysis of lead free and highly efficient intrinsic $CH_3NH_3SNI_3$ based solar cell with suitable transport layers. Results in Optics. 2023;**13**:100517. Available from: https://linkinghub.elsevier.com/retrieve/pii/S2666950123001694

[38] Sahoo D, Karan AK, Mallick Z, Manik NB. Synthesis and complex impedance analysis of nano cubic $CH_3NH_3SnI_3$ perovskite for the development of optoelectronic lead-free Schottky diode. Materials Science in Semiconductor Processing. 2023;**155**:107253. Available from: https://linkinghub.elsevier.com/retrieve/pii/S136980012200779X

[39] Sahoo D, Karan AK, Manik NB. Influence of SWCNT on the electrical behavior of an environmentally friendly $CH_3NH_3SnI_3$ perovskite-based optoelectronic Schottky device. ACS Applied Electronic Materials. 2023;**5**(4):2203-2214. Available from: https://pubs.acs.org/doi/10.1021/acsaelm.3c00086

[40] Yamada K, Nose S, Umehara T, Okuda T, Ichiba S. 81Br NQR and 119Sn Mössbauer Study for $MSnBr_3$ (M=Cs and CH_3NH_3). Bulletin of the Chemical Society of Japan. 1988;**61**(12):4265-4268. Available from: https://academic.oup.com/bcsj/article/61/12/4265/7345207

[41] Onoda-Yamamuro N, Matsuo T, Suga H. Thermal, electric, and dielectric properties of $CH_3NH_3SnBr_3$ at low temperatures. The Journal of Chemical Thermodynamics. 1991;**23**(10):987-999. Available from: https://www.sciencedirect.com/science/article/pii/S002196140580179X

[42] Chiarella F, Ferro P, Licci F, Barra M, Biasiucci M, Cassinese A, et al. Preparation and transport properties of hybrid organic–inorganic $CH_3NH_3SnBr_3$ films. Applied Physics A: Materials Science & Processing. 2006;**86**(1):89-93. Available from: http://link.springer.com/10.1007/s00339-006-3733-7

[43] Swainson I, Chi L, Her JH, Cranswick L, Stephens P, Winkler B, et al. Orientational ordering, tilting and lone-pair activity in the perovskite methylammonium tin bromide, $CH_3NH_3SnBr_3$. Acta Crystallographica Section B: Structural Science. 2010;**66**(4):422-429. Available from: https://scripts.iucr.org/cgi-bin/paper?S0108768110014734

[44] Jung MC, Raga SR, Qi Y. Properties and solar cell applications of Pb-free perovskite films formed by vapor deposition. RSC Advances. 2016;**6**(4):2819-2825. Available from: http://xlink.rsc.org/?DOI=C5RA21291J

[45] Xu H, Yuan H, Duan J, Zhao Y, Jiao Z, Tang Q. Lead-free $CH_3NH_3SnBr_3$-xIx perovskite quantum dots for mesoscopic solar cell applications. Electrochimica Acta. 2018;**282**:807-812. Available from: https://linkinghub.elsevier.com/retrieve/pii/S0013468618311897

[46] Yamada K, Fujise K, Hino S, Yamane Y, Nakagama T. Characterization of Sn(II)-based perovskites by XRD, DTA, NQR and ^{119}Sn NMR for photovoltaic applications. Chemistry Letters. 2019;**48**(7):749-752. Available from: https://academic.oup.com/chemlett/article/48/7/749-752/7339096

[47] Chiarella F, Zappettini A, Licci F, Borriello I, Cantele G, Ninno D, et al. Combined experimental and theoretical investigation of optical, structural, and electronic properties of $CH_3NH_3SnX_3$ thin films (X = Cl, Br). Physical Review B. 2008;**77**(4):045129. Available from: https://link.aps.org/doi/10.1103/PhysRevB.77.045129

[48] Borriello I, Cantele G, Ninno D. *Ab initio* investigation of hybrid organic-inorganic perovskites based on tin halides. Physical Review B. 2008;**77**(23). Available from: https://link.aps.org/doi/10.1103/PhysRevB.77.235214

[49] Mosca R, Ferro P, Besagni T, Calestani D, Chiarella F, Licci F. Effect of humidity on the a.c. impedance of $CH_3NH_3SnCl_3$ hybrid films. Applied Physics A: Materials Science & Processing. 2011;**104**(4):1181-1187. Available from: http://link.springer.com/10.1007/s00339-011-6407-z

[50] Feng J, Xiao B. Effective masses and electronic and optical properties of nontoxic $MASnX_3$ (X = Cl, Br, and I) perovskite structures as solar cell absorber: A theoretical study using HSE06. The Journal of Physical Chemistry C. 2014;**118**(34):19655-19660. Available from: http://pubs.acs.org/doi/10.1021/jp506498k

[51] Wang LZ, Zhao YQ, Liu B, Wu LJ, Cai MQ. First-Principles Study of Photovoltaics and Carrier Mobility for Non-toxic Halide Perovskite $CH_3NH_3SnCl_3$: Theoretical Prediction - Physical Chemistry Chemical Physics. Vol. 18. China: RSC Publishing; 2024. pp. 22188-22195. Available from: https://pubs.rsc.org/en/content/articlelanding/2016/cp/c6cp03605h#!

[52] Wang L, Ou T, Wang K, Xiao G, Gao C, Zou B. Pressure-induced structural evolution, optical and electronic transitions of nontoxic organometal halide perovskite-based methylammonium tin chloride, 111. Applied Physics Letters. 2017;(23):233901. DOI: 10.1063/1.5004186

[53] Tsai CM, Mohanta N, Wang CY, Lin YP, Yang YW, Wang CL, et al. Formation of stable tin perovskites Co-crystallized with three halides for carbon-based mesoscopic Lead-free perovskite solar cells. Angewandte Chemie. 2017;**129**(44):14007-14011. Available from: http://doi.wiley.com/10.1002/ange.201707037

[54] Rahul SPK, Parvaz M, Ahmed S, Sonker RK, Bhattacharya B, et al. Less toxic tin incorporated perovskite solar cell using polymer electrolyte processed in the air. Optik. 2018;**169**: 166-171. Available from: https://linkinghub.elsevier.com/retrieve/pii/S0030402618307137

[55] Moyez SA, Roy S. Thermal engineering of lead-free nanostructured $CH_3NH_3SnCl_3$ perovskite material for thin-film solar cell. Journal of Nanoparticle Research. 2018;**20**(1):5. Available from: http://link.springer.com/10.1007/s11051-017-4108-z

[56] Kumar AA, Kumar A, Quamara JK. Lanthanum containing barium

stannate nanoparticles synthesized by cetyltriammonium bromide assisted wet chemistry route for application in perovskite solar cell. Materials Today: Proceedings. 2019;**17**:1487-1493. Available from: https://www.sciencedirect.com/science/article/pii/S2214785319315238

[57] Kumavat SR, Sonvane Y, Gupta SK. Structural, optical, transport, and solar cell properties of 2D halide perovskite MAZX3 (Z = Pb, Sn, and X = Cl, Br, I). Journal of Applied Physics. 2020;**128**:114304. Available from: https://aip.scitation.org/doi/10.1063/5.0016331?af=R&feed=most-recent&utm_source=researcher_app&utm_medium=referral&utm_campaign=RESR_MRKT_Researcher_inbound

[58] (17) (PDF) Synthesis Process and Structural Properties of Single and Mixed Hallide Sn-based Pervoskite Materials in Photovoltaic Application. 2024. Available from: https://www.researchgate.net/publication/365344873_Synthesis_Process_and_Structural_Properties_of_Single_and_Mixed_Hallide_Sn-based_Pervoskite_Materials_in_Photovoltaic_Application

[59] Halide double perovskite-based efficient mechanical energy harvester and storage devices for self-charging power unit - ScienceDirect. 2024. Available from: https://www.sciencedirect.com/science/article/abs/pii/S2211285522012265

[60] Chigari SS, Vidyasagar CC, Vishwanath CC, Faniband S. Ultrasonic radiation assisted synthesis of $(CH_3NH_3)_2CuCl_4$, $CH_3NH_3PbCl_3$, and $CH_3NH_3SnCl_3$ perovskites for energy application. Journal of hazardous materials. Advances. 2023;**12**:100368. Available from: https://www.sciencedirect.com/science/article/pii/S2772416623001390

Chapter 5

Photovoltaic Parameters Affecting the Efficiency and Stability of Perovskite Photovoltaic Devices

Asha Chauhan, Anjali Oudhia and Ashok Kumar Shrivastav

Abstract

Perovskite photovoltaic devices (PVDs) have emerged as excellent futuristic photovoltaic energy-harvesting material in the past few years with a remarkable efficiency of over 25%. The intense scientific research in the field of perovskite photovoltaic technologies further enables their goal of effective commercialization. The device's efficiency strongly relies on the optical, morphological and electrical properties of the emerged charge collective and perovskite layer. This chapter critically articulates the advancements made by the choice of charge collecting layer to get the optimum PVDs. Further, the challenges of lead toxicity and the possible replacement with green non-toxic perovskite will also be discussed. The influence of photovoltaic parameters such as the thickness, bandgap, electron affinity, mobility, dielectric permittivity, density of states, donor and acceptor density and defect level on the device's performance and stability are introduced in this chapter. The chapter also analyzes the impact of series resistance caused by various factors such as bulk resistance, circuit connection and contacts, uneven morphology, defects and voids/vacancies and trap states at grain boundaries of the perovskite layer. This chapter focuses on elucidating the theoretical mechanisms to optimize the photovoltaic performance of the device by tuning the photovoltaic properties and material modeling of different stoichiometry.

Keywords: photovoltaic device, perovskite, photovoltaic parameters, optimization, energy and efficiency

1. Introduction

1.1 Worldwide energy demand

Energy is essential to life and key to industrial development for the advancement of the economic and social well-being of the world population. The daily growth of the world population accelerated energy usage and consumption, resulting in an energy crisis and environmental pollution. The earth takes millions of years to fertilize and store the fuels. In the long history of human existence, the consumption of fossil fuels is a pulse action of relatively short duration owing to their exhausting supply and environmental impact. Fossil fuels (oil, coal, and natural gas) have major impacts on the environment and living things because their use is the emission of greenhouse gases, which leads to global warming and climate change. The new era of technology and industrialization is increasing the consumption and demand of energy all over the world. Hence, the requirement for alternate-sustainable and eco-friendly energy resources has risen, which can fulfill the energy hunger of the growing human population and replace fossil fuel sources. Due to the rising worldwide energy demand, new terms like the efficiency of energy, vulnerability in the supply, and the effects on the environment are becoming very popular.

The global energy consumption is estimated to be 580 million terajoules due to rapid economic expansion, population, urbanization, and industrialization. Environmental Impact Assessment (EIA) expects that the global energy demand to be increased by 47% in the next 30 years driven by recent technology vogues like machine learning, big data analytics, communication infrastructure like 5G, and the standard of living, centralized and standardized digital models. Petroleum Planning and Analysis Cell (PPAC), a unit of the federal oil ministry, released a report stating that the fuel demand in India could rise to 214.5 million tons in the next 30 years. The demand for gasoline, diesel, aviation fuel, petcoke (a better-burning alternative to coal), liquefied petroleum gas, and cooking fuel is estimated to grow by 10%. The huge energy demand in the various sector from households to industry could be met only by harnessing all available resources of energy. Also, to avail the natural energy resources for a longer period and for next-generation mankind, their conservation without affecting the productivity and targets is a must duty of the inhabitations of this planet. It is a must to develop energy conservation techniques, methods, and energy-saving projects and training. These methods are as simple as installing more insulation in the house, getting used to public transportation, and energy audit services like diesel use analysis, green workplace lightening, heat loss, efficiency, and planning for improvement in production. The oversized, overloaded, and overheated machines eat the energy and compress the machine's efficiency. The audit of energy usage, housekeeping, and maintenance of machines can reduce the consumption of energy by up to 5%.

The worldwide rapid consumption of fossil fuel resources is a subject of concern and urgently requires an alternate, trustworthy, and non-fossil fuel resource of energy to meet the energy demands for the immediate future and the upcoming generation. Renewable or sustainable non-fossil fuels such as the sun, wind, and water are derived from naturally replenished sources or processes of the earth. The most common sustainable energy resources are as follows: wind energy, solar energy, hydroelectric energy, geothermal energy, bioenergy, nuclear energy, hydrogen energy, tidal energy, and wave energy. Out of these energy resources, nuclear and solar energy stand out as the brightest long-range and traditional sources of energy that can fulfill the continuous energy demand. Nuclear energy originated from the process of putting light atoms together is called the nuclear fusion process. The earth has an abundant source of nuclear materials which makes it an alternative resource. Unfortunately, the radioactive waste from nuclear power plants and reactions has a great impact on the environment and climate. Constructing, monitoring, and maintaining nuclear power plants requires well-trained and skilled laborers. The need for huge investment and advanced technology are other drawbacks of nuclear power plants. Also, a nuclear accident can cause major damage to the health issues of living beings and the environment. From there, nuclear energy cannot be reasonably expected as the ultimate energy solution shortly.

The issues of nuclear pollution could be resolved by inexhaustible and renewable solar energy. Solar power is expanding exponentially throughout the world since it is a form of pollution-free light radiation and heat. A range of cost-effective technology to harness solar energy has been developed, such as solar photovoltaic, solar heating, and solar thermal electricity. Solar energy is used in water heating, heating buildings, power generation, photovoltaic conversion, solar cookers and ovens, solar drying of agricultural products, solar wind energy, etc. Modern solar technologies diversify the energy source and save money by large-scale electricity generation from solar energy.

1.2 Worldwide solar energy availability

Solar energy is an abundant energy source on Earth. Solar energy is a freely available and environmentally clean source of light like fossil fuels and nuclear energy. Solar radiation originates from nuclear fusion reactions that occur inside the core of the sun and the reaction produces the energy that keeps the sun hot. Sun is a massive nuclear reactor and 4 million of hydrogen is being converted into helium. In nuclear fusion, 4 hydrogens are converted into one helium, and some fraction of mass is released in the form of energy. The produced energy from the core of the sun is transmitted to the earth through electromagnetic radiation in the form of energy packets or quanta (photons). Sun strikes the earth continuously with a total of 173,000 terawatts of solar energy at a rate of 5E+24 Joule per year, which is more than 1000 times the energy produced by the world. Out of this huge number, only 6000 TWh of energy is converted into various forms of renewable energy (**Figure 1**) [1].

The energy produced per hour could meet the requirement of world energy for 1 year. However, due to atmospheric attenuation, not all part of the sunlight reaches the earth's surface. About 30% of sunlight is reflected in space and only 70% is absorbed

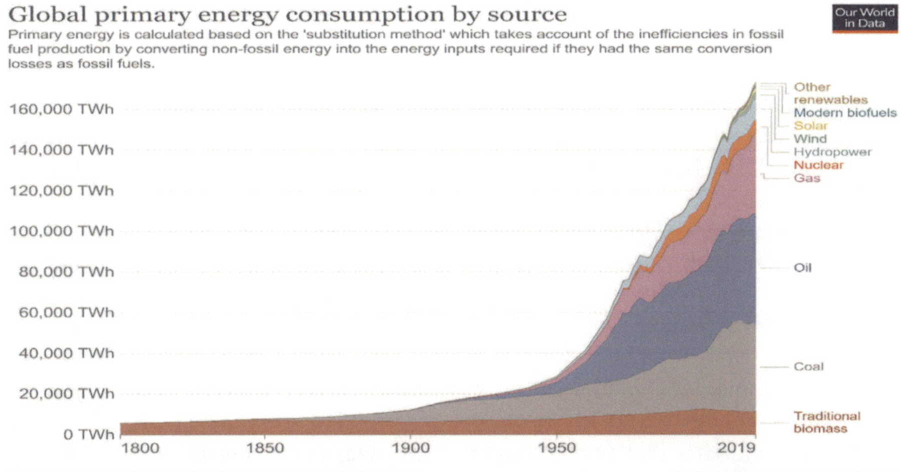

Figure 1.
Global direct primary energy consumption from 1800 until 2019. (Source: https://ourworldindata.org/grapher/global-primaryenergy?country=~OWID_WRL) [1].

on the earth's surface. About 16 out of 70% of absorbed light on the earth's surface is absorbed by dust and water vapor. The other 3% is absorbed by the clouds. Currently, solar energy is not being used as the primary resource of fuel energy; however, intense research work is required to harvest and store more energy from the sun to meet the high-energy demand of the world.

To harness solar energy, various sophisticated and modern photovoltaic (PV) technologies have been developed. Upon exposure to sunlight, the material produces electrical energy known as the photovoltaic effect. A solar cell is a photovoltaic device that converts light energy into electrical energy. An assembly of individual PV cells mounted in a framework for the installation is called solar panels. A collection of solar panels is called a PV panel and a sequence of PV panels is termed an array. The array of PV systems supplies electricity to the electrical device and equipment. The PV panels usually convert 15–22% of solar energy into electrical energy. The installation of solar panels can make a significant difference to the planet. The world implanted 260 GW of solar power plants and it is expected that in the future, this number will increase dramatically. In the United States, 1 GW of solar power is utilized for residential purposes. According to the report of the International Energy Agency (IEA), the process of converting photon energy into light energy is 720 TWh and it is expected to grow up to 1940 TWh in the upcoming years.

Despite harnessing solar energy, some common issues are associated with its use. The solar cells have intermittency issues and solar energy is diluted. According to Stefan-Boltzmann, the available solar radiation flux at the sun's surface is 63E+6 W/m^2. But in the hottest place on the earth, its availability is 1373 W/m^2, which is significantly low for PV devices and technological utilization of solar energy. It means that during night time and cloudy or rainy weather, solar radiation is not continuously available to generate electrical energy. Also, the availability of solar energy varies widely at different locations and under various weather conditions. Consequently, a large area of production and storage of electrical energy is required for many optoelectronic and PV device applications and it increases the total operating cost of the device. Thus, the main challenge for solar energy in the twenty-first century is its storage. So, the stored energy can be used in the event of a power outage. The researcher needs to develop a cost-effective and pollution-free method to collect and store the energy obtained from solar energy. This will prolong its utilization and help maintain a balance between the generation of electricity and the energy demand.

Another major challenge of solar energy is to improve the light absorbance property of solar cell materials. To harvest more energy from the sun needs to be absorbed more photons by the light-harvesting material. So, the light-harvesting material of the solar device can produce a huge number of excitons to get unprecedented performance from the device. Thus, we are required to find new excellent quality materials to use in the solar device. Among these factors, it is also worth noting that solar panels are implemented in the outer atmosphere and the decomposition of the solar cells in the ambient environment (oxygen and moisture) is also a big problem. The sensitivity toward moisture deteriorates and limits the actual performance of the solar device. The degradation mechanism and decomposition of the device need to be explored at ground level and need to look for strategies suppressing the thermal stability. Thus, it is required to deal with a lot of challenges of the solar cell to consider solar energy as an alternative energy resource. Intense, innovative, and seminal scientific works are required to modify the worldwide energy picture in the upcoming future which will help in power generation, creating a cleaner world, and improving our daily lives. Solar energy can be utilized in many different ways as the sun creates two types of energy: (i) light and (ii) heat. These forms of solar energy can be harnessed by both nature and human beings from photosynthesis in plants to PV cells, solar power, heating water and food, heating buildings, solar lighting, and solar wind energy.

1.3 Solar cell

Solar cells are the device that converts light energy directly into electrical energy through the photovoltaic effect. This process was discovered as early as 1839 and the invention of the first solar cell was reported in 1954 in the Bell Laboratory. The first reported cell was produced on silicon (Si) wafers with a recorded efficiency of 6% using either monocrystalline or polycrystalline crystals. Since then, many solar cells based on CdTe, CdS, CIGS, GaAs, etc., have been reported. The most recent and powerful solar cells in current PV technology are polymer-based solar cells, dye-sensitized solar cells, Si-based solar cells, and perovskite solar cells. Si-based solar cell dominates the current PV market by achieving a high efficiency of 26% in parallel to a modest fabrication cost. The contemporary PV market is dominated by Si-based solar cells because they can achieve a high efficiency of 26% while maintaining a low fabrication cost. Concurrently, a single-junction GaAs-based solar cell achieved an efficiency of 29.3% [2]. The working mechanism and three basic attributes required to operate a solar cell are as follows (**Figure 2**).

i. The generation of excitons or electron-hole pairs upon exposure to light illumination.

ii. The separation of opposite charge carriers toward the respective charge transport layers.

iii. The extraction of dissociated charge carriers to an external circuit of the device.

The energy from the sun to the earth's surface is primarily electromagnetic radiation, which contains a spectral region from 0.3 to 3.0 μm. This spectral region corresponds to the wavelength ranging from ultraviolet to near-infrared. The radiation reaching the earth is scattered and absorbed and the intensity of absorbance depends on the incident angle of radiation. Depending on this latitude, the energy value at the

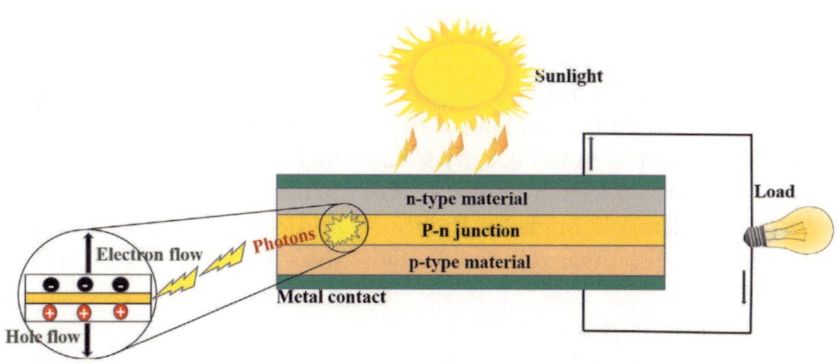

Figure 2.
Working principle of a solar cell.

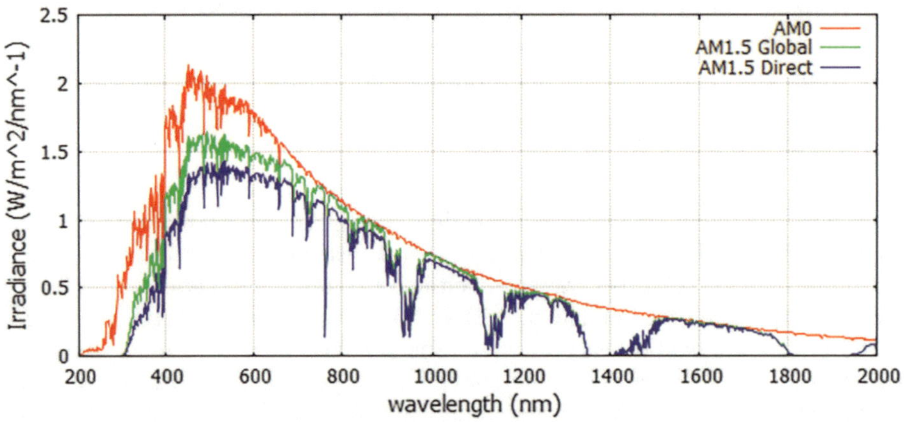

Figure 3.
Standard solar spectrum. (Source: https://www.solar-simulator.info/solar-simulator-match.html).

earth's surface is 1000 W/m^2. The power level outside the earth's surface is about 140 mW/cm^2, which is termed the air-mass-zero (AM0) spectrum. At sea level, with the sun at its zenith, the power level is reduced to 100 mW/cm^2, which is termed AM 1.5G solar radiation (**Figure 3**). The conversion of energy into electricity or spectral response by the solar cells is maximum at this spectral range. The spectral response is the ratio of incident energy to the total conversion of electricity by solar cells. PV technology's primary goal is to increase solar energy absorption in order to produce more power in a more cost-effective, environmentally friendly, and efficient manner.

1.4 Perovskite semiconductor

In 1839, Gustav Rose first discovered $CaTiO_3$, calcium titanium oxide perovskite mineral, in the Ural mountain and is called after Lev Perovski, a mineralogist. Broadly speaking, an assortment of substances termed perovskite crystal structure if it has a crystal structure that is comparable to that of $CaTiO_3$, which has the general formula ABO_3, which was first discovered in 1926 by Victor Moritz Goldschmidt while working on tolerance factor [3]. Goldschmidt defined the first description of the tolerance factor (t), which is an indicator of the stability and distortion of the crystal structure [3]. In a perovskite crystal structure, if a large number of stable metallic elements are present, then t is in the range of 0.75 to 1.

As one of the most abundant structural families, perovskites are found in an enormous number of compounds that have wide-ranging intriguing properties like ferroelectricity, superconductivity, charge ordering, magnetoresistance applications, and importance [4]. The first halide perovskite crystal structure based on cesium (Cs) and lead (Pb) was reported in $CsPbX_3$ by C.K. Moller in 1958 in his study of photoconductivity of the perovskite materials [5]. Tin (Sn)-based halide perovskite was first introduced as conducting channels for the thin-film field-effect transistor more than 20 years ago and in 2012, it was used as the light absorber layer in PSCs [6].

1.5 Properties of perovskites

Perovskite material showed outstanding photovoltaic properties like tunable bandgap, excellent thermal stability, long electron-hole diffusion length, high absorption coefficient, high charge carrier mobility and transportation, low processing temperature, lower binding energy, and easy processing steps. These basic characteristic properties of perovskite materials make them suitable for a wide range of applications in the field of PV technology. The family of metal halide perovskite crystals adopts a general chemical formula ABX_3. In which, A and B are two different size cations, often A atoms are larger than B atoms and X is an oxide or anion bonds with the other two cations. Additionally, A is monotonic and B is a divalent metal cation [7].

Halide perovskite is of two types: (i) all-inorganic halide perovskite and (ii) mixed or hybrid organic-inorganic halide perovskite. In pure halide perovskite, A refers to Cs^+, B refers to Pb^{2+}, Sn^{2+}, Ge^{2+}, etc., and X can be usually of halide group Cl^-, Br^-, I^-, etc. While in a mixed or organic-inorganic halide group of perovskite, A can be CH_3NH_3, MA^+, FA^+, etc., and X can be a mixture of Cl^-, Br^-, and I^-. In an ideal cubic perovskite crystal structure, "A" cations are positioned at (0, 0, 0) in the cube's corner and are 12-fold coordinated by halogen anions, whereas the halogen atoms "X" are located at position (1/2, 1/2, 0) in the cubic lattice's face center, and tetravalent "B" cations, on the other hand, are octahedral and occupy the location of the body's center (1/2, 1/2, 1/2) [8]. The oxide and halide perovskite can be found in different phases of crystal structure cubic, tetragonal, and rhombohedral depending on the tolerance factor and processing temperature (**Figures 4** and 5).

The tendency to form the perovskite and non-perovskite crystal phase depends on the formula of the tolerance factor, which is as follows [10]:

$$t = \frac{R_A + R_X}{\sqrt{2}(R_B + R_{X)}} \tag{1}$$

Where R_A = radius of A cation; R_B = radius of B cation; and R_x = radius of X anion.

In general, the tolerance factor (t) can lie between 0.89 and 1 for oxide perovskite and halide perovskite, it may vary from 0.89 to 1.11. The optimal cubic phase of

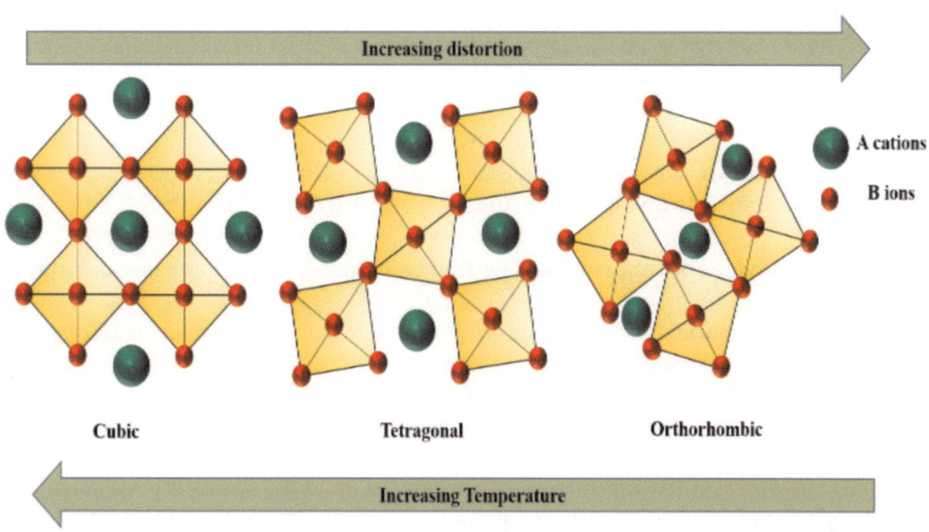

Figure 4.
The phase distortion from cubic to orthorhombic crystal phase V_s the temperature, 2D diagram of $CsPbBr_3$ perovskite material.

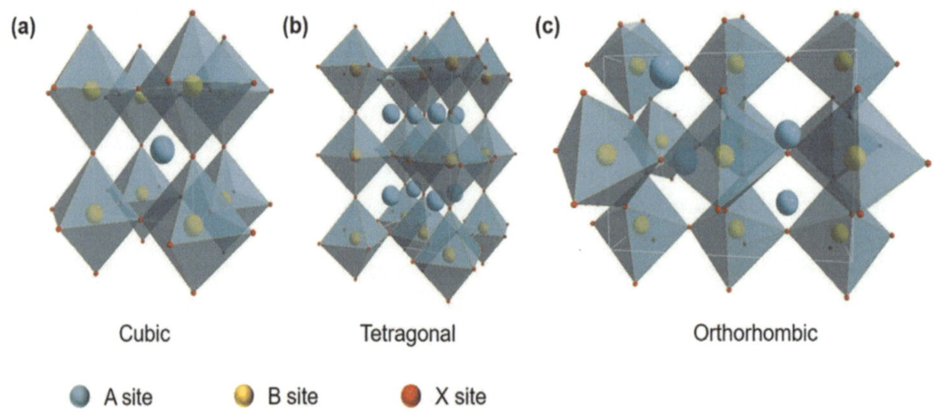

Figure 5.
The phase distortion from cubic to orthorhombic crystal phase 3D diagram of $CsPbBr_3$ perovskite [9].

perovskite is characterized by a t value of 0.8 to 1.0; bigger (>1) or lesser (<1) values of the t factor typically lead to non-perovskite structures [10]. The orthorhombic crystal structure is formed when t < 0.8, while a greater value of t > 1 gives a hexagonal phase. The ideal cubic structure tends to form a tolerance factor between 0.8 and 1. For a lower processing temperature, an orthorhombic crystal phase is obtained while a lower processing temperature gives a cubic crystal structure. The following trend of tolerance factor is defined for oxide perovskites, but it is still valid for halide groups of perovskites. Hence, the tolerance factor can be considered a reliable empirical index to predict the stability and preferentially formed stoichiometry of perovskite materials [10].

The direct bandgap perovskites have drawn much attention from researchers in photovoltaic applications from solar cells to light detectors and LEDs due to their excellent optical properties like broad absorption ranging across the whole visible spectrum, light absorption coefficient, and tunable bandgap which can be considered as one of the best and promising candidates to replace the Si technology.

To harvest solar energy, the light absorber layer of a PSC is the essential part. The high absorption coefficient of order 10^5 cm^{-1} and broadened absorption window covering ultraviolet to long-range spectrum at a lower thickness of the light absorber layer reduce the fabrication and experimental cost of the device in PSC. The thickness of the perovskite layer in PSC is relatively low in comparison with dye-sensitized solar cells. This may encourage the development and commercialization of PSC at a low cost.

According to Shockley and Queisser (SQ) limits or detailed balance calculation, the maximum power conversion efficiency from a single-junction solar cell is 30% for a single-junction solar cell with a bandgap of 1.1 eV of light-absorbing materials; however, the current calculation increased the efficiency to 33.16% at a bandgap of absorber layer 1.34 eV (**Figure 6**).

Despite this SQ calculation, the single-junction solar cell can hardly touch the upper limits. Other than the bandgap of the absorber layer, various factors like defect level, interfacial defects, doping concentration, vacancies, and interstitials limit the theoretical maximum efficiency. Out of 100% of solar energy, only 33% was converted into consumable electricity, and out of the remaining 67%, 47% was

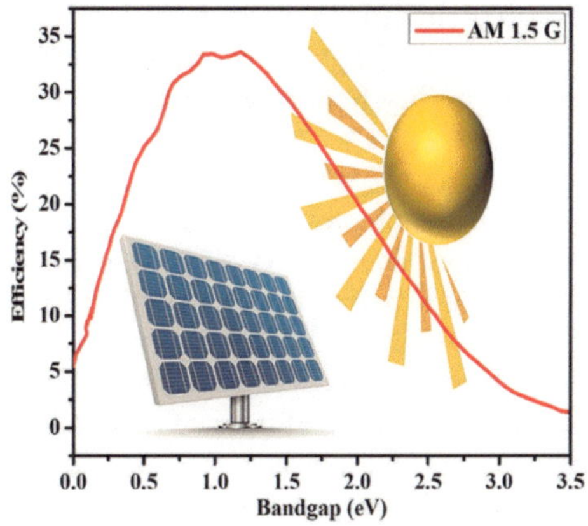

Figure 6.
Shockley and Queisser limits [11].

converted into heat, 18% was transmitted through the cell, and 2% was lost from the local recombination process of charge carriers.

The broad tunability of compositional choices upon the main three components of perovskite A, M, and X can lead to a variety of perovskites with different optical properties, which allow us substantial flexibility in choosing suitable perovskite material for specific optoelectronic applications (**Figure** 7) [12–14].

The tunable bandgap of perovskite materials enables efficient absorption of light and produces large excitons in a light-emitting gadget or photovoltaic device when exposed to light. The bandgap of perovskite materials can be easily adjusted by merely altering the halides' mixing ratio (Br-Cl, I-Br, etc.) [15]. The tunability of the bandgap of perovskite allows the creation of multicolor light-emitting devices and any specific required device. Hence, compositional engineering can give an easy path to obtain the desired optical property, which sufficiently enlarges the absorption window of perovskite material to utilize solar energy.

The perovskite material exhibits outstanding electrical properties such as conductivity, excitons binding energy as low as 2 meV [16], charge carrier mobility >100 $cm^2V^{-1}\ s^{-1}$, the diffusion length of electrons and holes exceeding 150 μm [17],

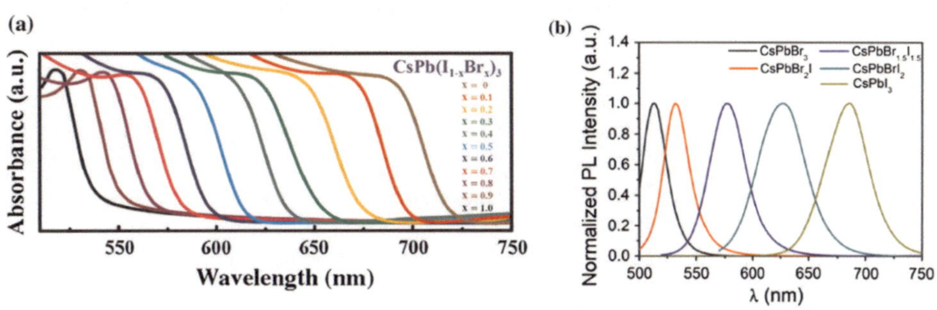

Figure 7.
Bandgap tuning of inorganic perovskite material by compositional engineering method (a) Absorbance spectra for inorganic perovskite films of different nanocomposites, $CsPb(I_{1-x}Br_x)_3$ Reprinted with permission from Ref. [12]. Copyright 2022, Elsevier B.V. (b) PL spectra of the $CsPbI_3$, $CsPbI_2Br$, $CsPbI_{1.5}Br_{1.5}$, $CsPbIBr_2$, $CsPbBr_3$ films. Reproduced with permission [13]. Copyright 2022, Elsevier B.V.

photoluminescence quantum yield (PLQY), low trap density of states 1E+10 cm^{-3} [18], and photoresponse as compared to the crystalline silicon [19]. Also, a unique property of this material is the ability to both act as a light-absorbing material and as a charge carrier transporter [20].

The mobility of charge carriers helps to determine the performance of the semi-conducting materials in optoelectronic applications. In general, the electron mobility is comparatively higher than the hole mobility since the effective mass of electrons is smaller than the holes. The holes function within the tightly bound valence band rather than the conduction band, which slows the movement of holes inside the potential well of semiconducting materials. The higher mobility >100 $cm^2V^{-1}\,s^{-1}$ of charge carriers in perovskite can be utilized in excellent carrier transportation character to improve the performance of perovskite-based cells [17, 21].

The operation mechanism of PSC depends on the fraction of excitons, which are thermally dissociated into free charge carrier electrons and holes upon light illumination on solar cells. PSCs produce a significant fraction of excitons within the perovskite layer depending on the energy bandgap. Perovskite materials reported a very low binding energy of 2 meV as compared to other semiconducting technology silicon [16]. In perovskite, generally two types of excitons are possible: (i) Frenkel excitons, which are tightly bound charge carriers that cannot move outside the unit cell, and (ii) Wannier excitons, which are weakly bound charge carriers free to move in the perovskite crystals. Wannier-Mott excitons are generally expected in the perovskites, which are often described in terms of the hydrogen model [22]. The binding energy depends on the temperature and it decreases at higher operating temperatures.

The average travel distance before the recombinations of charge carriers in perovskite crystals is significantly higher >150 μm and therefore the longer lifetimes of charge carriers [17]. The growth of perovskite microcrystals and rods provides a longer diffusion length of charge carriers. The lifespan and diffusion length are interconnected to each other as the higher diffusion length more the carrier's lifetime which is beneficial for the optoelectronic device. The diffusion length and lifetime of electrons and holes also depend on the defect density of the absorber layer; the higher defect level of the absorber layer limits the lifespan and diffusion length of charge carriers.

The acquisition of extreme thermal properties in perovskites has become more and more important because both the lifetime and performance of the device are majorly influenced by the temperature or temperature gradient [23]. Metal halide perovskite, $CsMX_3$, has excellent thermal stability, conductivity, and heat dissipation as compared to other organic perovskites (**Figure 8**) [23]. The various morphologies of perovskites

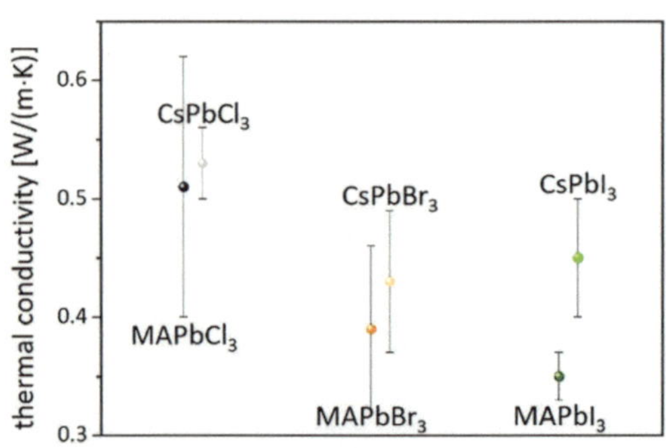

Figure 8.
Comparatively higher thermal conductivity in $CsMX_3$-based perovskites [23].

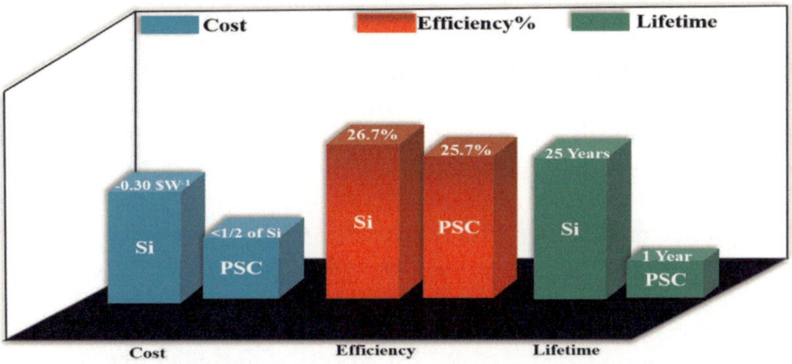

Figure 9.
Comparison of perovskite and silicon solar cells on golden triangle cost, efficiency, and lifetime. Si solar cells have a champion efficiency of 26.7% and a lifetime of over 25 years at a cost of around 0.3 $\$W^{-1}$ [26]. In comparison, PSC achieves an efficiency of 25.7% and a lifetime of 1 year at a cost of half of the silicon solar cell [25].

like micro-rods and the phase transformation to a stable phase reduce the rate of endothermic and exothermic reactions, which improve the self-thermal stability of the device [24].

The perovskites gained huge interest from experimenters due to the low cost of raw processing materials, which are half of silicon solar cells, higher efficiency up to 25.7% [25], and durability of around 1 year (**Figure 9**). However, currently, silicon solar cells achieve an efficiency of 26.7% and can operate over 25 years for the cost of 0.3 $\$W^{-1}$ [26]. Despite the higher efficiency of silicon solar cells, it is a very complex and lengthy process to extract Si from sand to Si wafer. Si technology is an expensive, high-energy process, generating a high level of waste material and is time-consuming. On the other hand, in a very convenient, easy processing way simpler, lower cost, lower embodied energy, and massively reduced environmental impact PSC can be produced and run. The large area role-to-role fabrication of inorganic $CsPbBr_3$, $CsPbI_3$, $CsPbI_2Br$, etc., perovskite thin film does not require any special processing environment. However, organic perovskites like $MAPbI_3$, $CH_3NH_3PbI_3$, $FASnI_3$ are very sensitive to moisture and oxygen present in the external environment and require inert (Nitrogen) processing conditions.

1.6 Perovskite solar cells

A perovskite solar cell (PSC) is a device that converts the photons of light from the sun into usable electricity.

A perovskite solar cell is made of three main parts:

i. Light-harvesting layer: a lead or tin-halide-based perovskite or semiconducting layer that acts as the light-harvesting active layer and converts the incident photons into electrons and holes [27]. Examples: all-inorganic perovskite—$CsPbBr_3$, $CsPbI_2Br$, $CsPbBr_2Cl$, $CsSnI_3$, alloy-based perovskite—$CsPb_{0.75}Sn_{0.25}I_2Br$, etc. The perovskite layer is sandwiched between the carrier collector layers as an active layer. The built-in electric field is generated in the perovskite and is due to the contact difference of potential between the ETL and HTL.

ii. Carrier collector layers: the p-type hole transport layer (HTL) and n-type electron transport layer (ETL) act as carrier collector layers that extract holes and electrons, respectively, from the perovskite layer. If the conduction band Ec and valence band Ev of both of the semiconducting materials lie below the Ec and Ev of the absorber layer, it will work as ETL and the opposite for HTL. Examples: ETLs-ZnO, TiO_2, GO, CdS and HTLs-Cu_2O, MoO_3, P3HT, etc.

iii. Metal contacts: The metallization layers on either side of the ETL and HTL layers act as electrical accessing electrodes and transfer the carriers to the circuit. If the work function of metal contact lies below and close to the conduction band of ETL or above the valence band for HTL, it will act as an electrode. Examples: C, Au, Ag, Cu, etc. The surface of PSC is coated with an anti-reflection coating to avoid the loss of incident light energy due to reflection. Also, to avoid external exposure of PSC to moisture and oxygen, PSC is encapsulated.

1.7 Working principle of perovskite solar cells

The working principle of PSC is based on the photovoltaic effect. The photogeneration of charge carriers in the light active layer is a result of the absorption of light radiation. When the photon hits the PSC, the active layer absorbs the solar energy and becomes a hotspot of the charge carriers. These charge carriers are loosely bound like metallic semiconductors, for example, silicon. Furthermore, the generated electrons dissociated toward the ETLs and holes dissociated toward the HTLs, which are extracted by the respective electrodes to the circuit, hence generating external electricity. The electricity is further amplified by using an inverter for different optoelectronic applications, **Figure 10**.

1.8 Thin-film perovskite solar cells

Thin-film perovskite solar cells are second-generation solar cells. In the thin-film solar cell, the thickness of the absorber, ETL, and HTL layer is in scale from a few nanometers (nm) to tens of micrometers (μm), which is much thinner than the rival technology: the first-generation crystalline silicon solar cell which has silicon wafers of

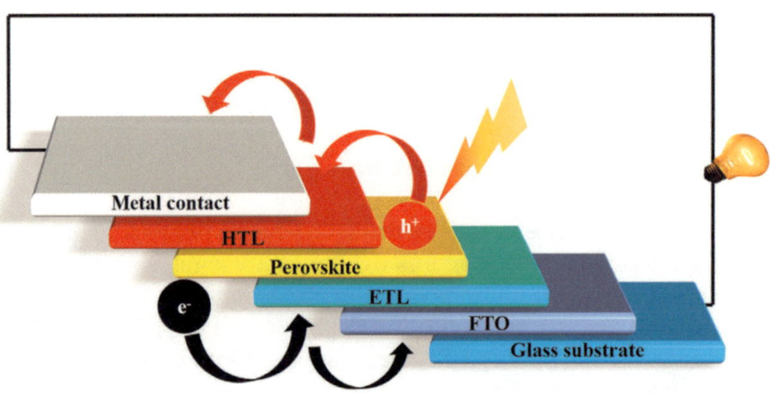

Figure 10.
Working principle of a PSC.

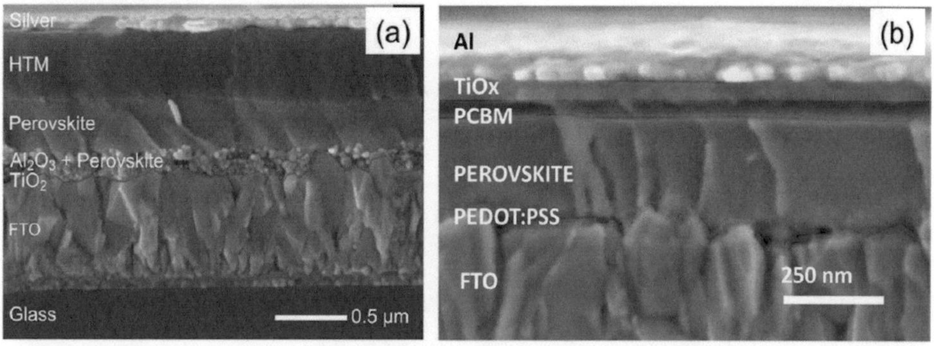

Figure 11.
(a) Cross-sectional SEM view of thin-film architecture-FTO/TiO_2/$Al2O_3$/Perovskite/HTM/Ag in the range μm. (b) cross-sectional view of a planar perovskite solar cell FTO/PEDOT: PSS/Perovskite/PCBM/TiO_x/Al in the range nm [28].

the thickness of 200 (μm) (**Figure 11**). The thin-film solar cell can be fabricated by depositing one or more thin layers on top of others. The powder of perovskite dissolves in the desired solvent to prepare a precursor solution for thin-film coating on top of a glass substrate, which is generally Fluorine-doped Tin oxide (FTO), in an extremely easy way. Further solution-coated wet film followed by drying and annealing to get the thin layer of solar cells. One of the most important advantages of thin-film solar cells is that they can be fabricated as flexible devices. The integration architectures of the flexible device with mass reduction have shown great potential in the field of wearable power supply due to flexibility, lightweight, portability, and compatibility without violating the actual function of the PSC [29].

2. Factors affecting the performance of perovskite solar cells

The performance of the PSC is affected by various photovoltaic parameters like bandgap of the absorber layer, type of ETL and HTL materials, and their thickness, interfacial recombination, series resistance, defect level of the absorber layer, and working temperature.

2.1 Bandgap

The active layer of the solar cell is a semiconductor that has a conduction band (CB) and valence band (VB). The gap between the CB and VB is termed as the bandgap of semiconducting materials. The conductors have no bandgap, semiconductors have bandgap smaller than 3 eV, and insulators have bandgap larger than 3 eV (**Figure 12**). The main reason for using semiconducting materials in solar cells is that sunlight excites the electrons from the CB to VB.

For insulating materials, solar energy is insufficient to excite the electrons to higher energy levels. There is an abundant source of semiconducting materials that have light-absorbing properties, which can be utilized in solar cells to generate power. In a semiconductor, photons with energy equal to or higher than the bandgap energy can lead to the absorption of light and produce excitons in the semiconductor [22]. Several mechanisms like the addition of impurity or compositional engineering of the

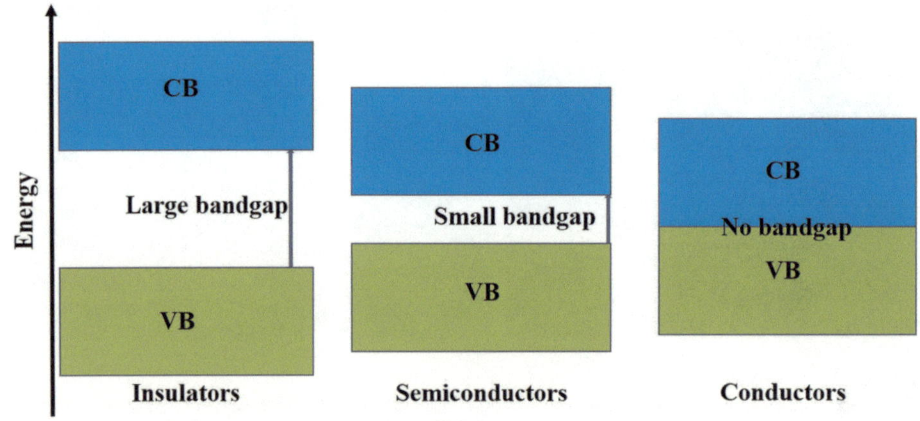

Figure 12.
Schematic diagrams of the energy band of insulator, semiconductor, and conductor.

base material can tune the bandgap and facilitate a higher PCE. Semiconductors are both direct and indirect materials and with the help of Tauc's equations, their bandgap can be calculated conveniently [30].

$$(\alpha h\upsilon)^n = k\left(h\upsilon - E_g\right) \tag{2}$$

Where h = Planck's constant; α = absorption coefficient; E_g = bandgap energy; k = proportionality constant; hυ = photon energy; n = 2, 1/2 for direct and indirect bandgap materials, respectively. In general, the relation between open-circuit voltage Voc and the band gap of the semiconducting material can be illustrated by the following equation:

$$V_{OC} = \frac{E_g}{q} - \frac{nKT}{q} log\left(\frac{J_{SC}}{J_0} + 1\right) \tag{3}$$

Where q = charge carriers; k = Boltzmann's constant; J_{SC} = short-circuit current; J_0 = reverse saturation current density; T = temperature; and n = ideality factor.

The Voc value of a solar cell will always be below the bandgap of the given materials, as the Voc is limited by the bandgap as shown in Eq. (3). The lower Voc in PSC is due to the recombination and non-radiative transitions of charge carriers. A Tin-based perovskite material overcomes the Voc loss by lowering the bandgap by adding the dopant of rubidium iodide (RbI) and potassium iodide (KI) in a tin-lead (SnPb)-based absorber and achieving a PCE of 21.04% (**Figure 13**) [9]. However, besides the bandgap, the stoichiometry, structures, and morphology differ from material to material, hence impacting the device's performance and possibilities for PV applications. A device with higher power conversion efficiency, low cost of raw materials, low consumption of processing energy, simple device production steps, and low price can be considered an excellent device.

2.2 Choice of ETL and HTL materials

Each functional layer of the PSC, absorber, ETL, and HTL layer plays a very important role in the photovoltaic performance of the device. When perovskite is

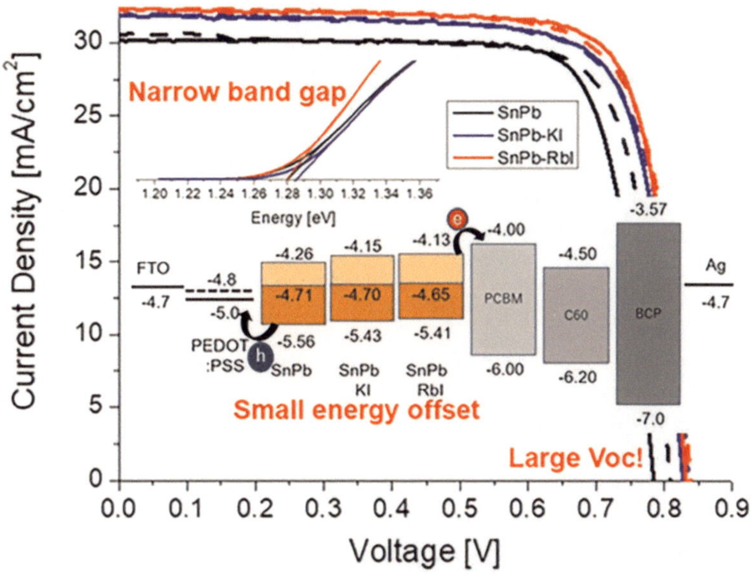

Figure 13.
A Tin-based PSC showed higher efficiency when tunning the bandgap. Reproduced with permission [9]. Copyright @ 2022, American Chemical Society.

structured with the various ETLs and HTLs materials, a mismatch between the energy level has risen due to the difference in the highest occupied molecular orbital (HOMO) and lowest unoccupied molecular orbit (LUMO) of functional materials. This discontinuity in the energy level of different materials gives the energy level diagram (**Figure 14**) [31]. The energy difference between the vacuum level and the conduction band minimum is provided by the electron affinity. The bandgap or the energy difference between the valence band maximum and the conduction band minimum is given by the symbol E_g. The energy bandgap and electron affinity values of each semiconducting material vary. The band diagram is positioned with the electron affinity of HTL, $CsPbBr_3$, and ETL layer (**Figure 14**). The energy level mismatch

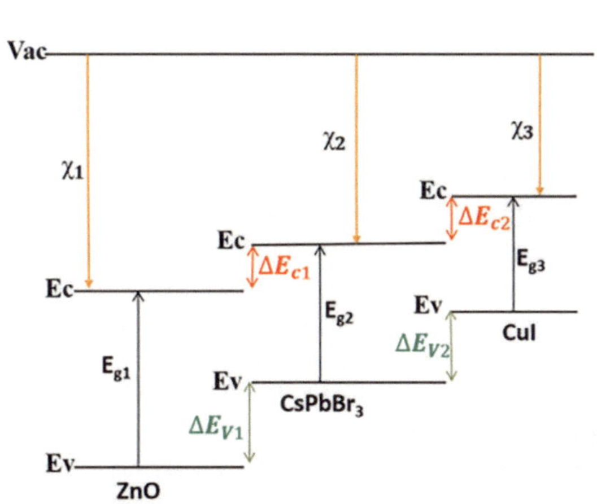

Figure 14.
Band diagram positioned with the energy bandgap and the electron affinity of HTL, $CsPbBr_3$, and ETL layer.

between the ETL/perovskite and perovskite/HTL causes a discontinuity at the interface, where the ETL-ZnO, perovskite-$CsPbBr_3$, and HTL-CuI come together. Note: in this diagram, we assume a constant Fermi energy level.

In **Figure 14**, x_1, x_2, and x_3 are the electron affinity of ETL-ZnO, perovskite-$CsPbBr_3$, and HTL-CuI, respectively. The conduction band and valence band energy are represented by Ec and Ev, respectively, while Vac is the vacuum level. E_{g1}, E_{g2}, and E_{g3} are the bandgaps of ETL, absorber, and HTL layers, respectively. The discontinuity is given by the following equations:

$$\Delta E_C = x_1 - x_2 \tag{4}$$

$$\Delta E_V = (x_1 + E_{g1}) - (x_2 + E_{g2}) \tag{5}$$

Where ΔE_c is the negative value representing the difference in the semiconductors' electron affinities. ΔE_V, which is a positive amount according to Anderson's law, is the sum of the differences in bandgaps and electron affinity [32, 33]. The discontinuities, or conduction band offset (CBO), at the $ZnO/CsPbBr_3$ and $CsPbBr_3/CuI$ interfaces, are represented by ΔE_{C1} and ΔE_{C2}, respectively, and are caused by the misalignment of the energy levels. The discontinuity's minimal optimal value, $\Delta E_c \sim (-0.2\ eV)$, the discontinuities at the $ZnO/CsPbBr_3$ and $CsPbBr_3/CuI$ interfaces are denoted by ΔE_{V1}, ΔE_{V2}, and VBO, respectively. 0.1 eV is the minimum value of ΔE_V. Positive conduction band offset occurs when ZnO's conduction band is lower than the $CsPbBr_3$ absorber layer's. It serves as a barrier to prevent electrons from moving backward from ZnO to the $CsPbBr_3$ layer. Conversely, a negative conduction band offset (CBO < 0) occurs when the ZnO conduction band is larger than the $CsPbBr_3$ layer's [34]. Because the backward flow of electrons cannot be sufficiently stopped by the negative CBO, charge carrier recombination takes over. In heterostructure solar cells with n-i-p, this conduction band offset occurs. Similar to a p-n junction solar cell, a depletion layer forms to cause band bending E_{F1}-E_{F2}, as both the hole and the electron move from the p- to the n-region and from the n- to the p-region, respectively. The n-i-p heterostructure device and p-n junction device are identical [35]. The device's permittivity, doping, and relative Fermi level all have a significant impact on the band bending value. As a result, we can adjust the CBO and VBO between the materials to raise the efficiency to a higher value by selecting different basic materials for the PSC.

Also, we are free to combine the different HTLs and ETLs materials with the perovskite layer, based on different factors like cost, nature of materials, alignment with the absorber layer, sintering temperature, lattice constant, etc. It is very easy to process an ETL-TiO_2 for PSCs, but its major drawback is requiring a high sintering temperature of 500°C and having very low mobility (0–0.1 $cm^2V^{-1}s^{-1}$) and low UV-stability [36]. To overcome this shortcoming, many other n-type metal oxide semiconductors like ETL- ZnO, SnO_2, GO, WO_3, $ZnSO_4$, CdS (**Figure 15**) can be replaced by TiO_2 [34, 37]. ZnO is an inorganic and low-cost material and it improved the stability of the device. Similar to the deposition of the perovskite layer, it is very easy to deposit the ETL layer on the glass electrodes. The choice of different materials may be appropriately aligned with the perovskite and give higher PCE from the device by reducing the total rate of recombination of charge carriers (**Figure 16(a)** and **(b)**). The higher CBO at the interface dramatically influences the current density V_s voltage curve (**Figure 16(d)**). Tunning the electron affinity of ETL and HTL materials to match the energy alignment with the absorber layer is another way to push the

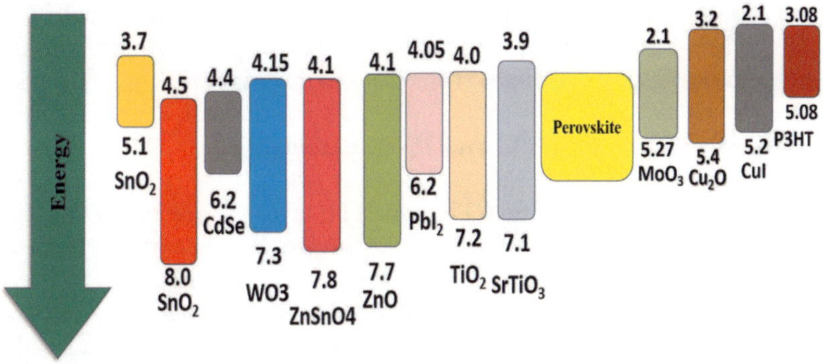

Figure 15.
Various choices of cost-effective ETL and HTL materials can be aligned with perovskites.

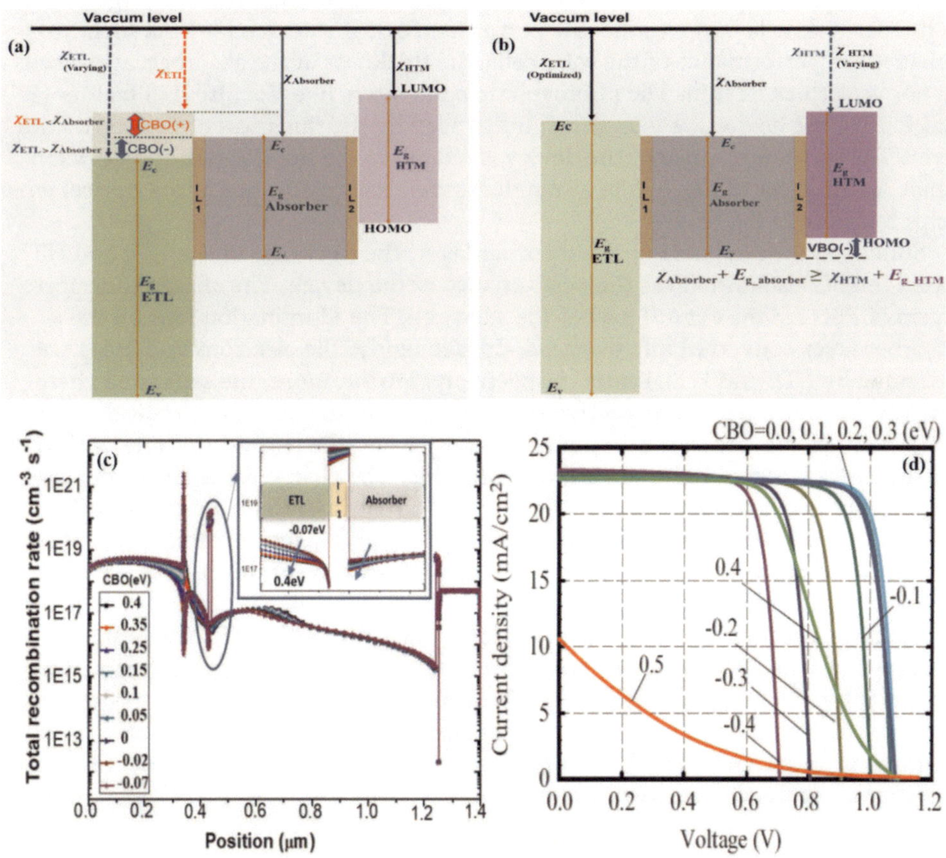

Figure 16.
(a, b) Formation of CBO and VBO at the interface ETL/perovskite and perovskite/HTL, respectively, (c) reduction in total recombination of charge carriers at interface ETL/perovskite and influence of CBO on (d) current density Vs voltage curve. Reproduced with permission [34]. Copyright 2020, Elsevier B.V.

efficiency of the device to a higher extent at a lower cost and with experimental wastage. Yassine et al. achieved a higher PCE of 25.21% by modulating the CBO of SnO_2 with the perovskite layer (**Figure 16(c)**). Similarly, in place of high-cost

Spiro-OMeTAD, the various p-type HTL materials like CuI, Cu_2O, MoO_3, P3HT, NiO, can be used for low-cost, high hole mobility and easy in-processing materials, which can fasten the extraction of the holes from the absorber layer.

2.3 Thickness of perovskite, ETL, and HTL materials

The thickness of the perovskite, ETL, and HTL layers is one of the most important parameters to enhance the performance of the device. Because of its effect on the electron and hole diffusion length of the charge carrier, the thickness of the absorber layer directly affects the solar cell's performance. As it is known, for an absorber layer with minimal thickness, the rate of light absorption is very low, which results in low photocurrent due to enhancement in carrier's recombination, hence reducing the device functionality. Conversely, a very high thickness of the absorber layer may not give an appropriate diffusion length to the charge carriers to travel toward the charge transport layers ETLs and HTLs; consequently, the recombination of charge carriers in the perovskite layer gives the low performance of the device [38]. As such, to enhance the performance of the solar cells, the thickness of the absorber layer needs to be not very thick or thin. The photovoltaic parameters like Voc, Jsc, Fill factors (FF), and PCE of the device are very much influenced by the thickness of the perovskite layer. The series resistance of the device is related to the bulkiness of the device; hence, a thin layer of perovskite also reduces the series resistance of the device, which helps to improve the PCE.

Similar to the thickness of the absorber layer, the thickness of the ETL and HTL layers in PSC also influences the performance of the device. The charge transport layers of PSC are the central part of the solar cell. The illumination falls on the absorber layer converted into excitons. Consequently, the electrons and holes are extracted by ETL and HTL layers, respectively. Furthermore, the separated charge carriers move toward the respective electrodes and in the external circuit through metal contact. However, it is important to fully cover the perovskite layer with HTLs for the dissociation of holes toward the electrode. The charge transport layers are responsible for the transportation of generated charge carriers from the absorber layer. Thus, the thickness of ETL and HTL layers can impact the production rate of charge carriers, their transportation, and the performance of the device. For a thin HTL layer, the holes travel a shorter distance toward the electrodes and reduce the probability of recombination; hence, the ideal thickness of the HTL layers should be in the range of 100–200 nm to fully cover the perovskite layer and for better performance of the device [39]. The thick ETL layer creates the bulkiness of the device, hence increasing the series resistance of the device. Therefore, it is critical to determine the proper ETL and HTL layer thickness in order to cover the crystallites without compromising the device's series resistance.

2.4 Defect density of perovskite

The photovoltaic actions like absorption of photons, carrier generations, and transportation process take place in the absorber layer. Hence, the absorber layer is the core of PSC. The quality and morphology of perovskites influences the performance of solar cells as they impact the lifetime of charge carriers and the rate of generation/recombination of electrons and holes [38]. The defect density of the absorber layer is a crucial parameter that needs to be a concern for the enhancement of the performance of the device. The defect in the perovskite layer majorly occurred due to poor morphology and non-uniform coverage of the perovskite on the glass substrate leading to the formation of vacancy and voids/pinholes. The defect in the absorber layer causes the direct linkage of perovskite with the ETL and HTL layer and enhances the recombination rate of electrons and holes that limit the overall performance of the device [40]. The defects in perovskite cause electrical instability in the device. Due to the uneven surface and morphology of the perovskite layer, defect levels like interstitial, dislocation, Schottky, and Frenkel defects can exist in between the bandgap of the material, which can eat and trap the electrons that result in non-radiative transitions that highly influence the PCE of the device [41, 42]. If the activation energy is sufficiently small, then trap electrons are likely to be emitted, back to CB (or VB) by phonon absorption. On the other hand, if the activation energy is large, then it is more likely that trapped electrons (or holes) will annihilate or recombine with an opposite charge carrier before it can be emitted [43]. The recombination with the opposite charge carriers can be non-radiative transitions that are assisted by phonon emission [43]. This is an important loss mechanism in PSCs and the rate of recombination is determined by the Shockley-Read-Hall (SRH) mechanism [44].

$$R_{RSH} = \frac{np - n_i^2}{\tau\left(p + n + 2n_i \cosh\left(\frac{E_i - E_t}{kT}\right)\right)} \tag{6}$$

$$L = \sqrt{(D \times \tau)} \tag{7}$$

$$\tau = \frac{1}{\sigma \times N_t \times v_{th}} \tag{8}$$

Where E_t = defect energy level; L = diffusion length of charge carriers; N_t = defect density; σ = carriers capture; v_{th} = thermal velocity; D = diffusion length; τ = lifetime of charge carriers; and n_i = electron concentration.

Furthermore, as illustrated in **Figure 17**, the departure from an ideal lattice structure is referred to as a defect that can occur in the perovskite crystal structure in a variety of ways, including vacancy, interstitial, dislocation, grain boundaries, Frenkel,

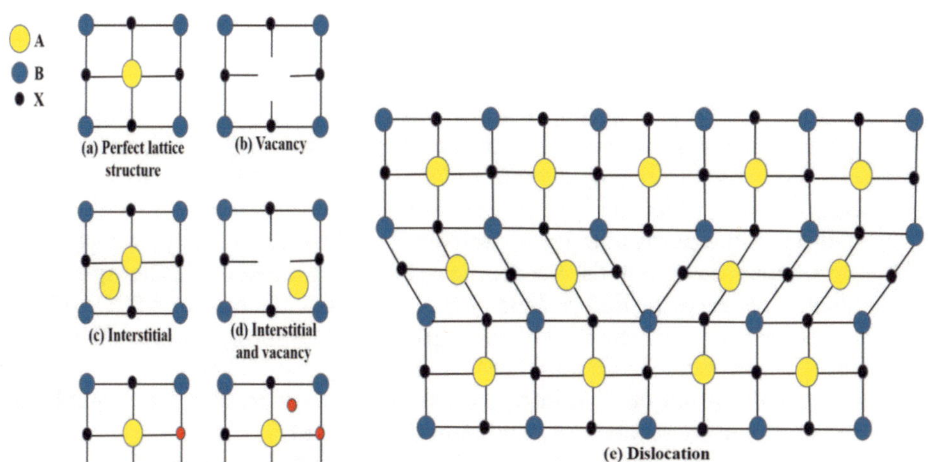

Figure 17.
Types of defects that can exist in perovskite material.

and Schottky defects [43, 45, 46]. These flaws affect exciton extraction, cause instability and non-radiative recombination, and prevent PSCs from being used commercially [46, 47]. Additionally, the literature review demonstrated that the following successful techniques can be used to experimentally reduce the absorber layer's defect density and increase the device's efficiency:

i. The device's stability, defect concentration, and performance are all greatly influenced by the thin perovskite layer deposition technique [46, 48]. Numerous factors, including stoichiometry balance, compound mixing technique, annealing temperature, grain size, solvent, and contaminants from the synthesis environment (such as oxygen and moisture), influence the concentration of defects in the absorber layer [49]. Numerous deposition techniques, doctor blading, chemical bath deposition (CBD), sequential vacuum deposition, including chemical vapor deposition (CVD), and spin-coating (one-step and two-step deposition), have been used for the synthesis of the perovskite layer for large area role-to-role fabrication were employed to create the perovskite layer for large-scale role-to-role manufacturing [46, 50–52]. Enhanced efficiency in the CVD process and one-step spin-coating deposition, compared to the two-step deposition approach, was demonstrated by Azat Akbulatov et al. Additionally, the step deposition method produces a quick nucleation process of perovskite film with a decreased root mean square (RMS) roughness, which results in a smoother, less imperfect, and pin-hole-free PSC light active layer [46, 53].

ii. Increasing the perovskite layer's grain size results in a low trap density of states, increased thermal stability, and less charge carrier recombination, all of which enhance the device's PCE [54, 55]. Moreover, the processing temperature and crystallization rate have a significant impact on grain size; large grains can be produced by gradually lowering the temperature [46, 47, 56–58].

iii. The thermally treated perovskite film, indicating a defect-free perovskite surface in PSC, has a lifespan of charge carrier recombination longer than 7 μs, [54].

iv. Perovskite thin-film doping and compositional engineering techniques can greatly reduce the status of the trap of the material, improving device efficiency by passivating defects and lengthening the charge carrier recombination lifetime [46, 59]. To increase PSC efficiency, the defect density of the perovskite layer can be reduced using a variety of deposition parameters and methods. According to Shoulong Chen et al., the AgBr doping quickly reduced the defect density from 5.059E + 15 cm^{-3} to 3.509E+15, improving the performance of the AgBr-doped $CsPb_{0.99}Ag_{0.01}Br_3$ PSC in comparison with the undoped cell. By lowering the VTFL [46], the addition of a suitable amount of AgBr to the CsPbBr3 film decreased the defect level in the photoluminescence (PL) emission (**Figure 18**).

The value of trap or defect density is given by the formula [59]:

$$n_{trap} = \frac{2\varepsilon\varepsilon_0 V_{TFL}}{qL^2} \tag{9}$$

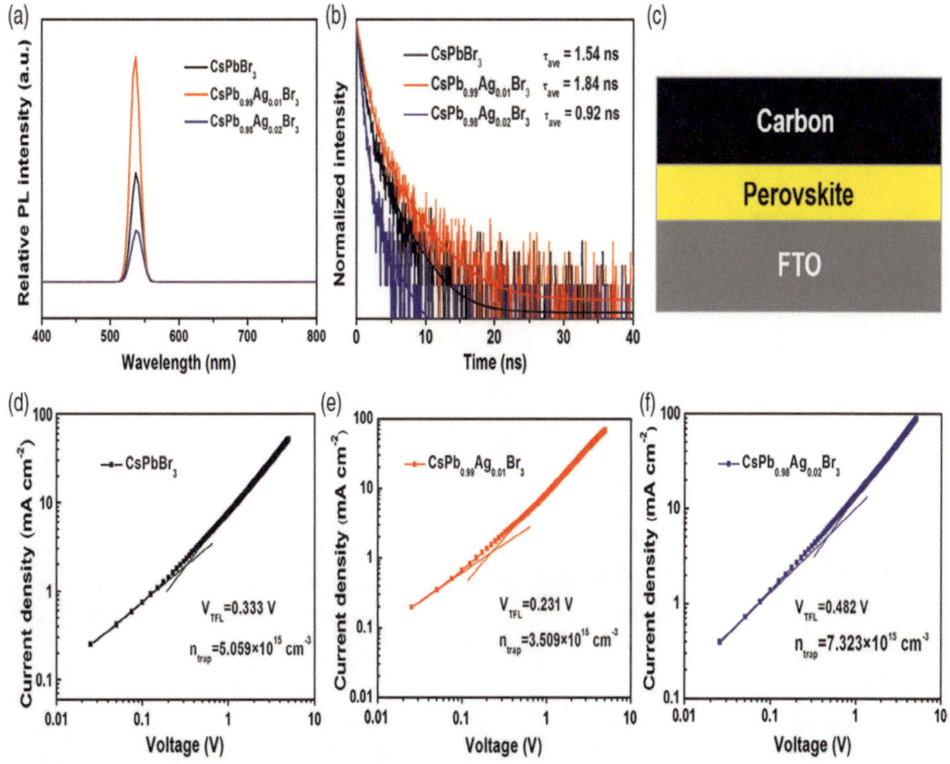

Figure 18.
(a) PL and (b) TRPL spectra for different $CsPb_{1-x}Ag_xBr_3$ (x = 0–0.02) films (c) Dark J-V measurement device structure: FTO/perovskite/carbon (d–f) Dark J-V curves characterization in different perovskite films [59].

Where ε_0 = vacuum permittivity; ε = dielectric constant; L = thickness of the film; V_{TFL} = trap-filled limited voltage.

It is also worth noting that excessing doping may increase the defect level. Hence, the defect level of the light-absorbing layer should be lower as much as possible.

2.5 Series resistance of the device

The performance of the device is significantly affected by the series resistance of the device and the series resistance existence in the device is majorly due to the non-uniform morphology of film, circuit connections, bulk resistance, front and back metal contacts, and defects in perovskite layer [34, 60]. The ETL layer was not covered completely because of the uneven and inadequate quality perovskite layer, which affects the ETL/perovskite interface and increases the contact resistance of the device [61]. The series resistance influences the performance of PSCs. The series resistance of the devices has very little impact on the Voc of the PSC. The major impact of the series resistance is on the FF and PCE of the device [62]. However, excessively high values of series resistance reduce the short-circuit current density (Jsc) of the device. Series resistance does not affect the solar cell at Voc since the overall current flows through the cell. However, near the Voc, the current density V_s voltage (J-V) curve is strongly affected by the series resistance (**Figure 19**).

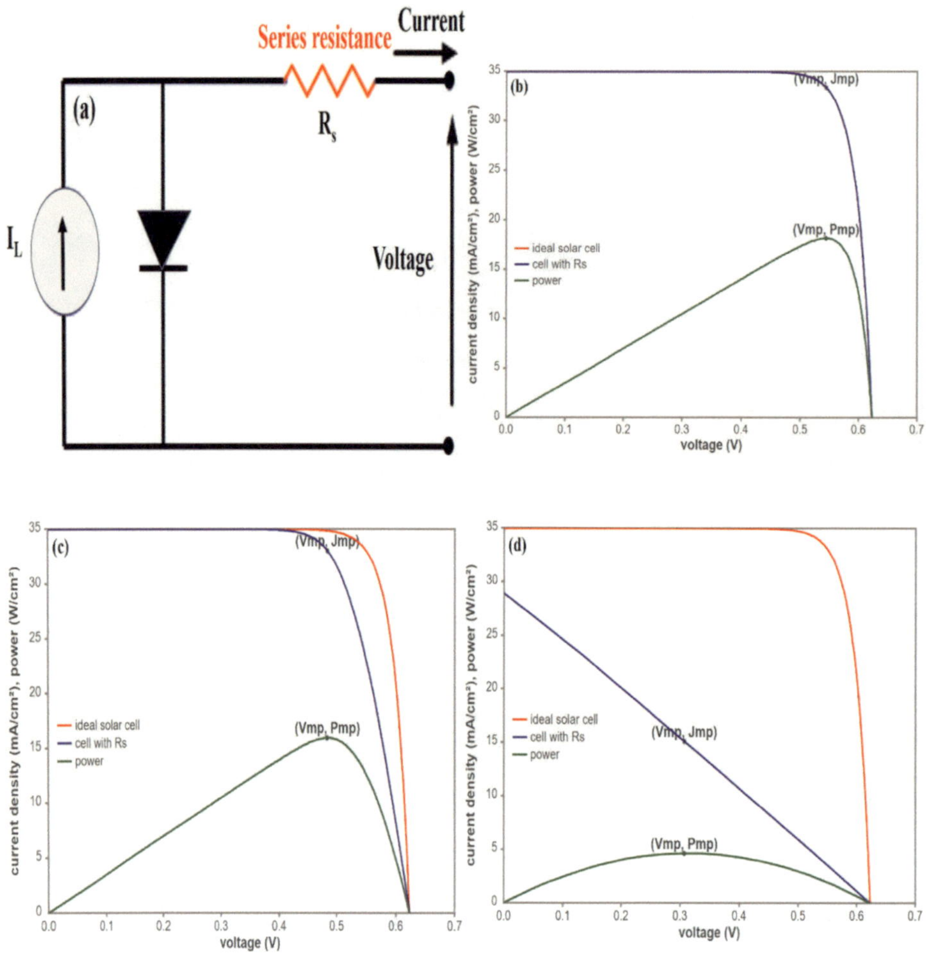

Figure 19.
(a) Equivalent circuit and impact on J-V curve at series resistance (b) 0 ohm.cm² (c) 2 ohm.cm² (d) 20 ohm.cm². (Source: https://www.pveducation.org/pvcdrom/solar-cell-operation/seriesresistance#:~:text=Series%20resistance%20in%20a%20solar,top%20and%20rear%20metal%20contacts).

The straightforward method to estimate the series resistance from the solar cell is to find the slope of the J-V curve at the Voc. The following formulas connect the Voc to Jsc using the device's diode equivalent circuit model [63]:

$$I_{SC} = I_o\left(exp\left(\frac{qV_{OC}}{e^{nKT}}\right) - 1\right) \tag{10}$$

$$I_{SC} = I_L - I_o\left(exp\left(\frac{qV_{OC}}{e^{nKT}}\right) - 1\right) - \frac{V_{oc} + I_{sc} \bullet R_s}{R_{SH}} \tag{11}$$

Where I_{SC} = short-circuit current; T = temperature; I_L = light-induced current; n = ideality factor; k = Boltzmann's constant; R_s = series resistance; and R_{SH} = shunt resistance.

We can see from the preceding calculations that an increase in series resistance has a major effect on the device's Jsc, which lowers the PCE and FF. To achieve the maximum PCE from the device, the series resistance of the device should be lower as much as possible. We must investigate and explore the configuration and use of the raw materials for the PSC, which have naturally lower series resistance like HTL-free configuration and carbon (C). HTL-free configuration of the cell lowers the bulkiness and contacts of the device, hence reducing the series resistance of the device. As a result, the device's defect level, film shape, series resistance, and film quality all contributed significantly to its efficiency.

2.6 Operating temperature

The operating temperature has a great impact on the performance of the PSCs. Perovskite materials are very sensitive to temperature and since the solar cells are implanted in the outdoor environment, it is very important to study how temperature can affect the device. The photovoltaic cells are operated under direct sunlight and therefore by the sun's heating, the cell temperature can be higher than 45°C, which is higher than the normal environment temperature [64]. Given the synthesis process of the PSCs, perovskites show three crystal phases, α, β, and γ, depending on the processing and crystallization temperature [64]. When processed at a higher temperature, the perovskite starts to decompose to constitute perovskite. However, sometimes the temperature-dependent processing method like the slow-lowering temperature method may be beneficial for the effective growth and crystallization of perovskite.

The stability of PSCs at the higher operating temperature is a critical challenge in the field of commercialization of PSCs. The higher working temperature breaks the bonds between molecules and improves carrier recombination. The unnecessary loss of charge carriers by recombination limits the actual potential of the solar cells. As seen from Eq. (3), the higher temperature significantly influences the J-V curve (**Figure 20**). Although degradation and recombination occurred at higher operating temperatures, sometimes a slightly higher temperature than the room temperature can increase the performance of the device by increasing the thermal concentrations [65].

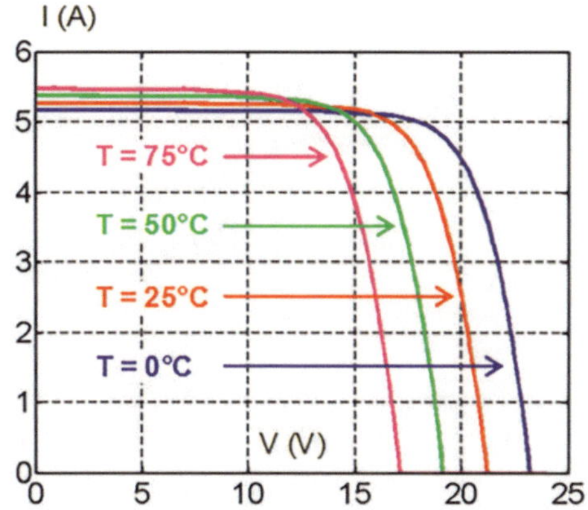

Figure 20.
Impact of working temperature on J-V curve. (Source: https://www.dekloo.net/projets/photovoltaique-autoconsommation/caracteristiques-panneau-photovoltaique/689).

However, the thermal stability can be significantly improved by the morphological variant of perovskite crystals. The morphology of large-scale perovskite crystals will not decompose easily in higher working temperatures as the rate of thermal decomposition, an endothermic and exothermic reaction is completely suppressed by the large morphology of crystals and reduces the self-heating property of perovskites [24]. In large-scale crystals, longer and more stable bonds between the molecules are formed, which facilely decompose under the higher operating temperature and reduce the rate of decomposition [58]. The long-range morphology like micro-rods is formed by stacking of crystallographic planes; hence, heat may not penetrate and break the compact arrangement of crystal planes which efficiently increases the self-life and thermal stability of the perovskites and reduces the degradation [66]. This strategy of tuning the size and scale of perovskites can also improve the thermal stability of the device.

3. All-inorganic metal halide ($CsMX_3$) perovskite solar cells

Nowadays, all-inorganic metal halide perovskite ($CsMX_3$, where M = Cs, Sn, X = Cl, Br, I) semiconductors have drawn increased attention in the development of PSCs, because of superior thermal, electronic, and photovoltaic properties [67–70]. Several studies revealed that 10–11% of all-inorganic lead halide PSCs performed as indicated [71]. However, more groundbreaking research and efforts are needed to raise the power conversion efficiency (PCE) to 25% [34]. The excellent intrinsic stability toward moisture is obtained by the cesium-based perovskite materials with an average efficiency of 10.26% [71]. The PCE of $CsMX_3$-based PSCs has sharply grown from 2.9 to 17.1% with much-improved stability since it was first published in 2015 [72]. The $CsMX_3$ has several morphologies including single-crystal and polycrystalline bulk films. $CsMX_3$ perovskite crystals can be prepared by solution-processed and bottom-up growth methods. $CsMX_3$-based perovskite shows excellent mobility and high PLQY; based on this property, $CsMX_3$ has been developed to apply to solar cells, photodetectors, and LEDs [73, 74].

In all-inorganic perovskites, $CsPbI_3$ is the best candidate for photovoltaic application with a cubic α phase and the narrowest bandgap of 1.73 eV can absorb a wide

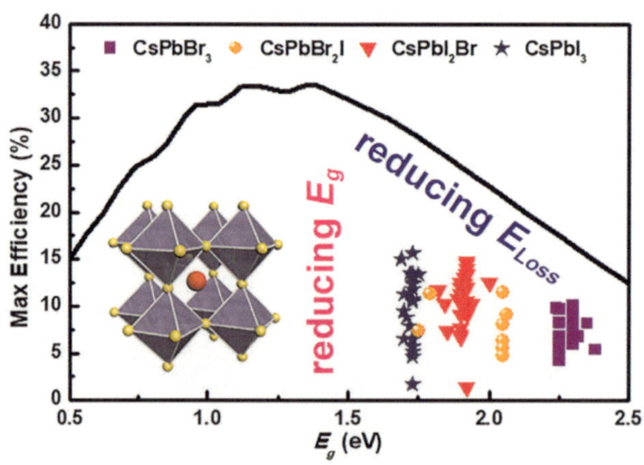

Figure 21.
The PCE of $CsMX_3$-based PSCs has abruptly increased to 17.1% [72].

range of solar radiation. When the bromide anion takes the place of the halide anion, the bandgap increases to 2.3 eV. The mixed-anion $CsPbI_2Br$ and $CsPbIBr_2$ exhibit a bandgap of 1.92 and 2.05 eV (**Figure 21**) [75]. $CsPbBr_2Cl$ is a new class of all-inorganic lead halide perovskite that needs to be explored. The polycrystalline nature of $CsPbBr_2Cl$ with a bandgap of 2.57 eV may be beneficial for the upcoming perovskite-based device. The $CsPbBr_2Cl$ has not been employed yet as a PSC due to the large bandgap [76]. However, these wide-bandgap perovskites have great potential in the field of multi-junction solar cells, LEDs, and photodetectors in combination with the low-bandgap perovskites [72].

4. Emerging Pb-free halide perovskite solar cells

The halide perovskites contain lead, which is a heavy metal that produces a toxic organic and inorganic compound that is hazardous to the environment. Therefore, it is important to investigate Pb-free light-absorbing metal halide materials, with elements such as tin (Sn), germanium (Ge), bismuth (Bi), and antimony (Sb), which can replace the Pb. For this reason, $CsSnX_3$, a Sn-based material free of Pb, can be an excellent alternative [77]. The oxidation of Sn^{2+} ions into Sn^{4+} ions, which is more stable than Sn^{2+} ions, causes Pb-free Sn-based perovskites $CsSnX_3$ to still have stability problems, but it also makes them extremely vulnerable to oxygen and moisture in the air [78].

Numerous studies on the synthesis process and solar applications of $CsSnX_3$-based perovskites have been published recently. The Cs_2SnX_6-based perovskite crystal phase was synthesized by Kaltzoglou et al. [79], who also reported a photovoltaic performance of 4.23% (**Table 1**). The PCE of Sn-based PSCs is still less than that of lead halide PSCs even if they do not have inherent toxicity. A Cs2SnI6-based perovskite structure with a bandgap of 1.48 eV and 1% efficiency was reported by Qiu et al. [81]. ZnO was used as the ETL, and Xiaofeng et al. [80] reported Cs2SnI6 PSC with an efficiency of 0.857%. They made use of HTL-P3HT and ETL-ZnO, the two inorganic absorber layers. $Cs_2SnCl_{6-x}Br$ photodetectors were reported by Jun et al. [87], while Cs_2SnCl_6 perovskite crystals were examined by Hongdan et al. [88]. According to Pujiarti et al. [82], the PCE of the $FTO/TiO_2/Cs_2SnI_6/Au$ PSC combination was 3.94%.

Currently, limited research work has been reported on Cs_2SnI_6-based perovskites, as this class of perovskite is still in the developing stage and a lot of investigation on

Device configuration	Voc (V)	Jsc (mA/cm^2)	FF (%)	PCE (%)
$FTO/ZnO/Cs_2SnI_6$/P3HT/Au [80]	0.52	3.20	51.50	0.857
$FTO/TiO_2/Cs_2SnI_6$/HTL/Ag [81]	0.51	5.41	32.0	1.0
$FTO/TiO_2/Cs_2SnI_6$/N719Dye/Pt [82]	0.67	16.3	61.8	6.8
$FTO/TiO_2/Cs_2SnI_6$/P3HT/Ag [83]	0.256	7.41	24	0.47
$FTO/CdS/Cs_2SnI_6$/C/Ag [84]	0.86	0.33	37	0.11
ITO/B-c-$CsSnI_3/Cs_2SnI_6$/Au [85]	0.63	6.1	28	1.1
FTO/PCBM/$CsGe_{0.5}Sn_{0.5}I_3$/Native oxide/Spiro-OMeTAD/Au [86]	18.41	0.63	61.3	7.11

Table 1.
Device performance comparison of PCE of Cs_2SnI_6-based PSCs.

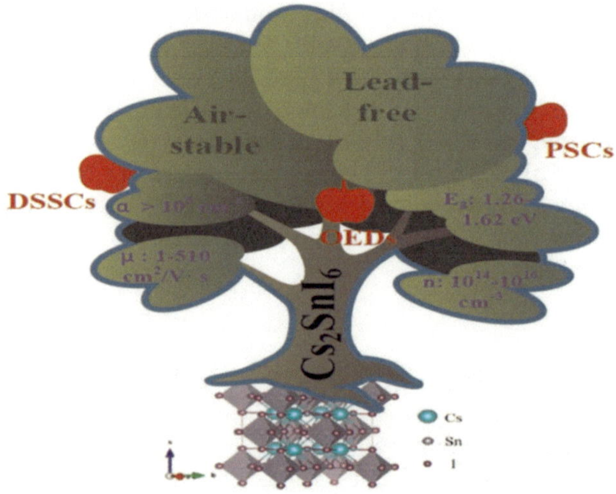

Figure 22.
Air stable and lead-free Cs_2SnI_6 is the next big thing in photovoltaics [89].

the structural, optical, and morphological is required to fully explore and understand the optoelectronic nature of the Cs_2SnI_6. The non-toxic nature and efficiency already accomplished in Cs_2SnI_6-based perovskite are not completed and fulfilled by this class of compounds. The various photovoltaic properties like bandgap and carrier mobility of Cs_2SnI_6-based perovskite material are shown in (**Figure 22**). The Cs_2SnI_6-based perovskite can be used in PSCs, LEDs, and dye-sensitized solar cells (DSSCs). The reported PCE by the Cs_2SnI_6-based PSCs is very low as compared to other classes of perovskites, hence various plenty rooms for these materials need to be filled by intense and seminal scientific and research work like compositional engineering, bandgap tuning, and synthesis methods to fascinating the highest efficiency and environmentally stable PSCs to fill the electricity requirements.

5. Challenges in the synthesis and commercialization of perovskite solar cells

Although PSCs have achieved rapid development in the field of photovoltaic technology in recent years, despite the PCE of PSCs touching the limit of Si-technology, the critical challenges in the synthesis process still limit their commercialization. However, to touch the S-Q limit, the efficiency of PSCs still needs to be improved. The fabrication process of PSCs mainly relied on a well-controlled environment, which is generally an inert gas (N_2)-filled glovebox. Inside a glovebox, a very small area of fabrication of PSCs is possible, which impedes its large area roll-to-roll ambient air scalable fabrications process. On the other hand, the fabrication of ETL, HTL, and perovskite layers under ambient conditions greatly affected the morphology, rate of nucleation, crystallization, and growth.

The poor voids/pinholes, non-uniform, and uneven spread on the surface of the glass substrate become a major challenge in the fabrication process. It is well believed that a good quality perovskite layer is responsible for the excellent photovoltaic performance of PSCs, which strongly rely on the formation dynamic of perovskite films. The synthesis process of the perovskite layer on the substrate also depends on the interaction between the perovskite precursor solution and the substrate. The perovskite layer faces several degradations and oxidation issues during the synthesis process in ambient air due to oxygen and moisture. The attack of oxygen molecules on perovskite decomposes into constituent perovskite molecules, which hindered the thermal stability of perovskites.

The doctor blade deposition and successive ionic layer adsorption and reaction (SILAR) method are one of the simplest large-scale and cost-effective film deposition techniques that have been widely used to fabricate perovskite thin film with uniformity and high crystallinity [90, 91]. It is also reported that the root mean square (RMS) of the roughness in the doctor-bladed (1.05 nm) and spin-coated (0.98 nm) $CsPbBr_3$ thin films are substantially equivalent [53]. However, the synthesis process by the doctor blade method is also attacked by oxygen and moisture leading to unsmooth perovskite film. The impurity doping, additive in the precursor solution, and compositional engineering may fill the voids/pinholes and provide smoother, compact, and uniform perovskite film [73].

In addition to this important area role in the fabrication of the perovskite layer, a certain amount of additive, such as nanoparticles with perovskite base material, is reported to improve the morphology, optical, and electrical properties of the perovskite light-harvesting layer, thereby improving the photovoltaic performances of PSCs [70, 92–97]. Nevertheless, organic additive materials like polyethylene oxide (PEO)-based precursor solution showed a high photoluminescence quantum yield (PLQY) of 60% when placed as a thin layer of perovskite $CsPbX_3$ [98]. Perovskite thin films are destroyed because PEO, an organic polymer material, is rapidly destroyed by moisture and necessitates intricate processing [99]. We require a stable additive that is resistant to environmental oxygen and moisture to enhance the quality of thin films. ZnO nanoparticles (NPs) support perovskite $CsPbX_3$ thin film for the photovoltaic performance of PSCs because ZnO is an inorganic and thermally stable material [73, 99]. In comparison to a pristine perovskite film, the NP-assisted films had fewer voids and a lower RMS roughness. The $CsPbBr_2Cl$ perovskite layer was stabilized by the solvent vapor treatment, which also increased crystallization and provided density [100–102]. To achieve homogeneous and dense perovskite films in PSCs, however, several other factors must also be taken into account throughout the production process, including processing temperature, environment, and crystallization rate [103]. This method of raising the perovskite film's quality considerably enhances its optical and morphological characteristics, which raises the device's PCE because efficiency depends on the morphology of the film.

6. Conclusion

This chapter emphasizes the need for global energy demand as well as the energy supplies that are currently available. In order to obtain the most effective photovoltaic device, this article systematically explains the perovskite energy harvesting semiconductor with intriguing photovoltaic properties. This article introduces the impact of photovoltaic parameters on the stability and performance of the device, including the thickness, bandgap, electron affinity, mobility, dielectric permittivity, density of states, donor and acceptor density, and defect level of different functional layers of photovoltaic device. The chapter will also address the difficulties surrounding the toxicity of lead and its potential substitution with environmentally friendly, non-toxic perovskite. This chapter further emphasizes understanding the theoretical mechanisms for modifying the device's photovoltaic features and modeling components that have different stoichiometry to optimally utilize its solar energy production.

Conflict of interest

There is no conflict of interest.

Author details

Asha Chauhan[1]*, Anjali Oudhia[2] and Ashok Kumar Shrivastav[1]

1 Department of Physics, National Institute of Technology Raipur, Chhattisgarh, India

2 Department of Physics, Government Nagarjuna Science P.G. College Raipur, Chhattisgarh, India

*Address all correspondence to: ashachauhan.phy@gmail.com

References

[1] Smil V. Global Solar Energy Consumption, Measured in Terawatt-Hours, Energy Transitions: Global and National Perspectives. BP Statistical Review of World Energy [Internet]. University of Oxford; 2017. Available from: https://ourworldindata.org/grapher/global-primary-energy?country=~OWID_WRL

[2] Polman A, Knight M, Garnett EC, Ehrier B, Sinke WC. Photovoltaic materials: Present efficiencies and future challenges. Science. 2016;**15**(352):6283. DOI: 10.1126/science.aad442

[3] Goldschmidt VM. Die Gesetze der Krystallochemie. Die Naturwissenschaften. 1926;**14**:477-485. DOI: 10.1007/BF01507527

[4] Cristina A. Crystal chemistry, stability and properties of interlanthanide perovskites: A review. Journal of the European Ceramic Society. 2017;**37**(2):427-440. DOI: 10.1016/j.jeurceramsoc.2016.08.041

[5] Jena AK, Kulkarni A, Miyasaka T. Halide perovskite photovoltaics: Background, status and future prospects, chemical reviews. ACS Chemical Reviews. 2019;**13**(119):3036-3103. DOI: 10.1021/acs.chemrev.8b00539

[6] Cao J, Yan F. Recent progress in tin-based perovskite solar cells. Energy & Environmental Science. 2021;**14**: 1286. DOI: 10.1016/j.mtener.2021.100891

[7] Yi Z, Ladi NH, Shai X, Li H, Shen Y, Wang M. Will organic-inorganic hybrid halide lead perovskites to be eliminated from optoelectronic applications? Nanoscale Advances. 2019;**1**:1276-1289. DOI: 10.1039/C8NA00416A

[8] Assirey EAR. Perovskite synthesis, properties, and their related biochemical and industrial application. Saudi Pharmaceutical Journal. 2019;**27**:817-829. DOI: 10.1016/j.jsps.2019.05.003

[9] Zhu Y, Deng K, Sun H, Gu B, Lu H, Cao F, et al. TiO_2 phase junction electron transport layer boosts the efficiency of planar perovskite solar cells. Advanced Science. 2018;**5**(1):1700471. DOI: 10.1002/advs.201700614

[10] Li Z, Yang M, Park JS, Wei SH, Berry JJ, Zhu K. Stabilizing perovskite structures by tunning tolerance factor: Formation of formamidinium and cesium lead iodide solid-states alloys. Chemistry of Materials. 2016;**28**(1): 284-292. DOI: 10.1021/acs.chemmater.5b04107

[11] Choudhury BD, Ibarra B, Cesano F, Mao Y, Huda MN, Chowdhury AR, et al. The photon absorber and interconnecting layers in multijunction organic solar cells. Solar Energy. 2020; **201**:28-44. DOI: 10.1016/j.solener.2020.02.035

[12] Jiang Y, Yuan J, Ni Y, Yang J, Wang Y, Jiu T, et al. Reduced-dimensional α-$CsPbX_3$ perovskites for efficient and stable photovoltaics. Joule. 2018;**2**(7):1356-1368. DOI: 10.1016/j.joule.2018.05.004

[13] Ghosh D, Ali MY, Chaudhary DK, Bhattacharyya S. Dependence of halide composition on the stability of highly efficient all-inorganic cesium lead halide perovskite quantum dot solar cells. Solar Energy Materials and Solar Cells. 2018; **185**:28-35. DOI: 10.1016/j.solmat.2018.05.002

[14] Ou Q, Bao X, Zhang Y, Shao H, Xing G, Li X, et al. Band structure

engineering in metal halide perovskite nanostructures for optoelectronic applications. Nano Materials Science. 2019;**1**:268-287. DOI: 10.1016/j.nanoms.2019.10.004

[15] Sutherland BR, Sargent EH. Perovskite photonic sources. Nature Photon. 2016;**10**:295-302. DOI: 10.1038/nphoton.2016.62

[16] Lin Q, Armin A, Nagiri RCR, Burn PL, Meredith P. Electro-optics of perovskite solar cells. Nature Photonics. 2014;**9**:106-112. DOI: 10.1038/nphoton.2014.284

[17] Dong Q, Fang Y, Shao Y, Mulligan P, Qiu J, Cao L, et al. Electron-hole diffusion lengths > 175 μm in solution-grown $CH_3NH_3PbI_3$ single crystals. Science. 2015;**347**:967-970. DOI: 10.1126/science.aaa5760

[18] Frolova LA, Dremova NN, Troshin PA. The chemical origin of the p-type and n-type doping effects in the hybrid methylammonium-lead iodide ($MAPbI_3$) perovskite solar cells. Chemical Communications. 2015;**51**: 14917-14920. DOI: 10.1039/C5CC05205J

[19] Shi D, Adinolfi V, Comin R, Yuan M, Alarousu E, Buin A, et al. Solar cells low trap-state density and long carrier diffusion in organolead trihalide perovskite single crystals. Science. 2015; **347**:519-522. DOI: 10.1126/science.aaa2725

[20] Motta C, Mellouhi FE, Sanvito S. Charge carrier mobility in hybrid halide perovskites. Scientific Reports. 2015;**5**: 12746. DOI: 10.48550/arXiv.1411.7904

[21] Fan GY, Jing DH, Nan LN, Lei Y, Yu ZC. Effect of carrier mobility on the performance of perovskite solar cells. Chinese Physics B. 2019;**28**(4):048802

[22] Baranowski M, Plochocka P. Excitons in metal-halide perovskites. Advanced Energy Materials. 2020; **10**(26):1903659. DOI: 10.1002/aenm.201903659

[23] Haeger T, Heiderhoff R, Riedl T. Thermal properties of metal-halide perovskites. Journal of Materials Chemistry C. 2020;**8**:14289-14311. DOI: 10.1039/D0TC03754K

[24] Mishra A, Ahmad Z, Touati F, Shakoor RA, Nazeeruddin MK. One-dimensional facile growth of $MAPbI_3$ perovskite micro-rods. RSC Advances. 2019;**9**:11589-11594. DOI: 10.1039/C9RA00200F

[25] NREL Efficiency chart. Available from: https://www.nrel.gov/pv/assets/pdfs/best-research-cell-efficiencies.20190802.pdf

[26] Meng L, You J, Yang Y. Addressing the stability issue of perovskite solar cells for commercial applications. Nature Communications. 2018;**9**:5265. DOI: 10.1038/s41467-018-07255-1

[27] Manser JS, Christians JA, Kamat PV. Intriguing optoelectronic properties of metal halide perovskites. Chemical Reviews. 2016;**116**(21):12956-13008. DOI: 10.1021/acs.chemrev.6b00136

[28] Sum TC, Mathews N. Advancements in perovskite solar cell: Photophysics behind the photovoltaics. Advanced Science. 2018;**5**(3):1700614. DOI: 10.1039/C4EE00673A

[29] Liang X, Ge C, Fang Q, Deng W, Dey S, Lin H, et al. Flexible perovskite solar cells: Progress and prospects. Frontiers in Materials. 2021;**8**:634353. DOI: 10.3389/fmats.2021.634353

[30] Borowski M. Perovskites: Structure, Properties and Uses (Chemical

Engineering Methods and Technology). UK ed. Nova Science Publishers, Inc; 30 Mar 2011. ISBN: 978-1-61668-525-6

[31] Andrea C, Ole H. What are the band alignments of $Cu_2ZnSn(S, Se)_4$ solar cells? Solar Energy Materials and Solar Cells. 2017;**169**:177-194. DOI: 10.1016/j.solmat.2017.05.008

[32] Sarunas S, Sergejus B, Mantas S, Gediminas U, Martynas K, Martynas V, et al. Phase transitions, screening, and dielectric response of $CsPbBr_3$. Journal of Materials Chemistry A. 2020;**8**: 14015-14022. DOI: 10.1039/D0TA04155F

[33] Chauhan A, Oudhia A, Shrivastav AK, Tirkey OS. Performance enhancement of an all-inorganic perovskite solar cell by using SCAPS and DFT extracted parameters of CuX (X = I, Cl, Br). Material Chemistry and Physics. 2023;**309**:128327. DOI: 10.1016/j.matchemphys.2023.128327

[34] Yassine R, Hamid EZ, Samrana K, Shahzada A. Energy level engineering of charge selective contact and halide perovskite by modulating band offset: Mechanic insights. Journal of Energy Chemistry. 2021;**54**:822-829. DOI: 10.1016/j.jechem.2020.06.030

[35] Guoqing T, Taotao C, Huan L, Wentao S, Yajing C, Jingling L, et al. Highly efficient hole extraction and stable all-bromine inorganic perovskite solar cells via derivative-phase gradient bandgap. Advanced Science News Solar RRL. 2019;**3**(5):1900030. DOI: 10.1002/solr.201900030

[36] Liu N, Chen MM, Yang H, Yang H, Ran M, Zhang C, et al. TiO_2/Mg-SnO_2 nanoparticle composite compact layer for enhancing the performance of perovskite solar cells. Optical Materials Express. 2020;**10**(1):157-169. DOI: 10.1364/OME.380354

[37] Mahmood K, Sarwar S, Mehran MT. Current status of electron transport layers in perovskite solar cells: Materials and properties. RSC Advances. 2017;7: 17044. DOI: 10.1039/C7RA00002B

[38] Yasin S, Zoubi TA, Moustafa M. Design and simulation of high-efficiency lead-free heterostructure perovskite solar cell using SCAPS-1D. Optik. 2021; **229**:166258. DOI: 10.1016/j.ijleo.2021.166258

[39] Bag A, Radhakrishnan R, Nekovei R, Jeyakumar R. Effect of the absorber layer, hole transport thickness and its doping density on the performance of perovskite solar cells by device simulation. Solar Energy. 2020;**196**: 177-182. DOI: 10.1016/j.solener.2019.12.014

[40] Ullah S, Liu P, Wang J, Yang P, Liu L, Yang SE, et al. Optimizing the working mechanism of the $CsPbBr_3$-based inorganic perovskite solar cells for enhanced efficiency. Solar Energy. 2020; **209**:79-84. DOI: 10.1016/j.solener.2019.12.014

[41] Chauhan A, Oudhia A. First-principle density functional theory-derived nonleaded $KSn_{1-x}Ge_xI_3$-based perovskite solar cells: A theoretical study. Energy Technology. 2023;**12**(2): 2300772. DOI: 10.1002/ente.202300772

[42] Lee YM, Maeng I, Park J, Song M, Yun JH, Jung MC, et al. Comprehensive understanding and controlling the defect structure: An effective approach for organic-inorganic hybrid perovskite-based solar-cell application. Journal of Solid State Chemistry. 2018;**205**:39-43. DOI: 10.3389/fenrg.2018.00128

[43] Ball JM, Petrozza A. Defects in perovskite-halides and their effects in solar cells. Nature Energy. 2016;**1**:16149. DOI: 10.1038/nenergy.2016.149

[44] Abdelaziz S, Zekry A, Shaker A, Abouelatta M. Investigating the performance of formamidinium tin-based perovskite solar cell by SCAPS device simulation. Optical Materials. 2020;**101**:109738. DOI: 10.1016/j.optmat.2020.109738

[45] Patel PK. Device simulation of highly efficient eco-friendly $CH_3NH_3SnI_3$ perovskite solar cell. Scientific Reports. 2021;**11**:3082. DOI: 10.1038/s41598-021-82817-w

[46] Chauhan A, Shrivastav AK, Oudhia A. Synthesis and characterization of ambient processed FTO/ZnO/$CsPbBr_2Cl$/C perovskite solar cell deposited by SILAR method. Optical Materials. 2022;**130**:112575. DOI: 10.1016/j.optmat.2022.112575

[47] Pratheek M, Abhinav T, Bhattacharya S, Chandra GK, Predeep P. Recent progress on defect passivation in perovskites for a solar cell application. Materials Science for Energy Technologies. 2021;**4**:282-289. DOI: 10.1016/j.mset.2021.07.003

[48] Lin CT, Rossi FD, Kim J, Baker J, Ngiam J, Xu B, et al. Evidence for surface defect passivation as the origin of the remarkable photostability of unencapsulated perovskite solar cells employing aminovaleric acid as a processing additive. Journal of Materials Chemistry A. 2019;**7**:3006-3011. DOI: 10.1039/C8TA11985F

[49] Akbulatov AF, Frolova L, Tsarev SA, Zhidkov IS, Luchkin S, Kurmaev EZ, et al. Film deposition techniques impact defect density and photostability of $MAPbI_3$ perovskite films. Journal of Physical Chemistry C. 2020;**124**(39):21378-21385. DOI: 10.1021/acs.jpcc.0c07673

[50] Tavakoli MM, Gu L, Gao Y, Reckmeier C, He J, Rogach AL, et al. Fabrication of efficient planar perovskite solar cells using a one-step chemical vapor deposition method. Scientific Reports. 2015;**5**:14083. DOI: 10.1038/srep14083

[51] Chen CW, Kang HW, Hsiao SY, Yang PF, Chiang KM, Lin HW. Efficient and uniform planar-type perovskite solar cells by simple sequential vacuum deposition. Advanced Materials. 2014;**26**(38):6647-6652. DOI: 10.1002/adma.201402461

[52] Leyden MR, Ono LK, Raga SR, Kato Y, Wang S, Qi Y. High-performance perovskite solar cells by hybrid chemical vapor deposition. Journal of Materials Chemistry. 2014;**2**:18742-18745. DOI: 10.1039/C4TA04385E

[53] Noor BM, Ahmed FH, Ammar EAK. Study of mechanical properties of Wollastonite filled epoxy functionally graded composite. International Journal of Mechanical Engineering and Technology. 2018;**9**:669-677

[54] Wang F, Bai S, Tress W, Hagfeldt A, Gao F. Defects engineering for high-performance perovskite solar cells. npj Flexible Electronics. 2018;**2**:22. DOI: 10.1038/s41528-018-0035-z

[55] Yang WS, Noh JH, Jeon NJ, Kim YC, Ryu S, Seo J, et al. High-performance photovoltaic perovskite layers fabricated through intramolecular exchange. Science. 2015;**348**(6240):1234-1237. DOI: 10.1126/science.aaa9272

[56] Tailor NK, Satapathi S. The impact of $Cs_3Bi_2Cl_9$ single crystal growth modality on its symmetry and morphology. Journal of Materials Research and Technology. 2020;**9**(4):7149-7157. DOI: 10.1016/j.jmrt.2020.04.097

[57] Gillespie C. How Does Temp Affect the Growth Rate of Crystals? Available from: https://sciencing.com/temp-affect-growth-rate-crystals-6318908.html 26 April 2018

[58] Niu G, Li W, Wang J. Progress of interface engineering in perovskite solar cells. Science China Materials. 2016;**59**: 728-742. DOI: 10.1007/s40843-016-5094-6

[59] Chen S, Liu X, Wang Z, Li W, Gu X, Lin J, et al. Defect passivation of $CsPbBr_3$ with AgBr for high-performance all-inorganic perovskite solar cells. Advanced Energy and Sustainability Research. 2016;**2**(6):2000099. DOI: 10.1002/aesr.202000099

[60] Semwal BS, Panwar NS. Dielectric properties of perovskite crystals. Bulletin of Materials Science. 1992;**15**:237. DOI: 10.1002/9781118935651.ch6

[61] Yang D, Yang R, Wang K, Wu C, Zhu X, Feng J, et al. High-efficiency planar-type perovskite solar cells with negligible hysteresis using EDTA-completed SnO_2. Nature Communications. 2018;**9**:3239. DOI: 10.1038/s41467-018-05760-x

[62] Mehrabian M, Afshar EN, Yousefzadeh SA. Simulating the thickness effect of the graphene oxide layer in $CsPbBr_3$-based solar cells. Materials Research Express. 2021;**8**: 35509. DOI: 10.1088/2053-1591/abf080

[63] Raj A, Kumar M, Bherwani H, Gupta A, Anshul A. Evidence of improved power conversion efficiency in lead-free $CsGeI_3$-based perovskite solar cell heterostructure via scaps simulation. Journal of Vacuum Science and Technology B. 2021;**39**:012401. DOI: 10.1116/6.0000718

[64] Mesquita I, Andrade L, Mendes A. Temperature impact on perovskite solar cells under operation. ChemSusChem. 2019;**12**(10):2186-2194. DOI: 10.1002/cssc.201802899

[65] Yadav P, Tripathi B, Lokhande M, Kumar M. Effect of temperature and concentration on commercial silicon module-based low-concentration photovoltaic system. Journal of Renewable and Sustainable Energy. 2013;**5**:013113. DOI: 10.1063/1.4790817

[66] Ding GQ, Shen WZ, Zheng MJ, Fan DH. Synthesis of ordered large-scale ZnO nanopore arrays. Applied Physics Letters. 2006;**88**(10):103106. DOI: 10.1063/1.2182025

[67] Son DY, Im JH, Kim HS, Park NG. 11% efficient perovskite solar cell based on ZnO nanorods: An effective charge collection system. The Journal of Physical Chemistry C. 2014;**118**(30): 16567-16573. DOI: 10.1021/jp412407j

[68] Zonglong Z, Yang B, Xiao L, Chu CC, Shihe Y, Alex K, et al. Enhanced efficiency and stability of inverted perovskite solar cells using highly crystalline SnO_2 nanocrystals as the robust electron-transporting layer. Advanced Materials. 2016;**28**(30): 6478-6484. DOI: 10.1002/adma.201600619

[69] Jie L, Fei G, Haoxu W, Juan L, Jiexuan J, Xin W, et al. Efficient planar $CsPbBr_3$ perovskite solar cells by dual-source vacuum evaporation. Solar Energy Materials and Solar Cells. 2018; **187**:1-8. DOI: 10.1016/j.solmat.2018.07.009

[70] Yongyi P, Yudiao C, Chunhua W, Chujun Z, Huayan X, Keqing H, et al. Fully doctor-bladed planar heterojunction perovskite solar cells under ambient condition. Organic Electronics. 2018;**58**:153-158. DOI: 10.1016/j.orgel.2018.04.020

[71] Haiwen Y, Yuanyuan Z, Jialong D, Yudi W, Xiya Y, Qunwei T. 10.26%-efficiency all-inorganic $CsPbBr_3$ perovskite solar cell by spectra engineering. Journal of Materials Chemistry A. 2018;**6**:24324. DOI: 10.1039/C8TA08900K

[72] Zhang J, Hodes G, Jin Z, Liu S. All-inorganic $CsPbX_3$ perovskite solar cells: Progress and prospects. Angewandte Chemie International Edition. 2019; **58**(44):15596-15618. DOI: 10.1002/anie.201901081

[73] Cunlong L, Zhigang Z, Ceng H, Zhiping H, Xiaosheng T, Juan D, et al. Enhanced random lasing emission from highly compact $CsPbBr_3$ perovskite thin films decorated by ZnO nanoparticles. Nano Energy. 2017;**40**:195-202. DOI: 10.1016/j.nanoen.2017.08.013

[74] Tongtong X, Xianfeng Y, Sunqi L, Junjian H, Yong L, Jinbo Y, et al. Highly stable $CsPbBr_3$ quantum dots coated with alkyl phosphate for white light-emitting diodes. Nanoscale. 2017;**9**(40): 15286-15290. DOI: 10.1039/C7NR04179A

[75] Deng J, Li J, Yang Z, Wang M. All-inorganic lead halide perovskite: A promising choice for photovoltaics and detectors. Journal of Materials Chemistry. 2019;**5**:1-68. DOI: 10.1039/C9TC04164H

[76] Paul T, Chatterjee BK, Maiti S, Besra N, Thakur S, Sarkar S, et al. Efficient blue emission from ambient processed all-inorganic $CsPbBr_2Cl$ perovskite cubes. AIP Conference Proceedings. 2018;**1942**:120026. DOI: 10.1063/1.5029066

[77] Lee B, Stoumpos CC, Zhou N, Hao F, Malliakas C, Yeh CY, et al. Air-stable molecular semiconducting iodosalts for solar cell applications: Cs_2SnI_6 as a hole conductor. Journal of the American Chemical Society. 2014;**136**(43): 15379-15385. DOI: 10.1021/ja508464w

[78] Xing G, Kumar MH, Chong WK, Liu X, Cai Y, Ding H, et al. Solution-processed tin-based perovskite for near-infrared lasing. Advanced Materials. 2016;**28**:8191-8196. DOI: 10.1002/adma.201601418

[79] Kaltzoglou A, Antoniadou M, Kontos AG, Stoumpos CC, Perganti D, Siranidi E, et al. Optical-vibrational properties of the Cs_2SnX_6 (X = Cl, Br, I) defect perovskites and hole-transport efficiency in dye-sensitized solar cells. Journal of Physical Chemistry C. 2016; **120**:11777-11785. DOI: 10.1021/acs.jpcc.6b02175

[80] Xiaofeng Q, Yanan J, Hailiang Z, Zhiwen Q, Shuai Y, Ping W, et al. Lead-free mesoscopic Cs_2SnI_6 perovskite solar cells using different nanostructured ZnO nanorods as electron transport layers. Physica Status Solidi RRL. 2016; **10**(8):587-591. DOI: 10.1002/pssr.201600166

[81] Qiu X, Cao B, Yuan S, Chen X, Qiu Z, Jiang Y, et al. From unstable $CsSnI_3$ to air-stable Cs_2SnI_6: A lead-free perovskite solar cell light absorber with bandgap of 1.48 eV and high absorption coefficient. Solar Energy Materials and Solar Cells. 2017;**159**:227-234. DOI: 10.1016/j.solmat.2016.09.022

[82] Pujiarti H, Wulandari P, Hidayat R. Photocurrent enhancement by incorporation of air-stable Cs_2SnI_6 perovskite in the dye-sensitized solar cell. Journal of Physics: Conference Series. 2019;**1245**:012066. DOI: 10.1088/1742-6596/1245/1/012066

[83] Jiang Y, Zhang H, Qiu X, Cao B. The air and thermal stabilities of lead-free perovskite variant Cs_2SnI_6 powder.

Materials Letters. 2017;**199**:50-52. DOI: 10.1016/j.matlet.2017.04.046

[84] Suazo FJA, Shaji S, Avellaneda D, Martínez JAA, Krishnan B. Solar cell using spray cast Cs_2SnI_6 perovskite thin films on chemical bath deposited CdS yielding high open-circuit voltage. Journal of Solar Energy. 2020;**207**:486. DOI: 10.1016/j.solener.2020.06.101

[85] Lee B, Krenselewski A, Baik SI, Seidmana DN, Chang RPH. Solution processing of air-stable molecular semiconducting iodosalts, Cs_2SnI_6-XBrX, for potential solar cell applications. Energy & Fuels. 2017;**1**:710. DOI: 10.1039/C7SE00100B

[86] Liu Y, Gao W, Ran C, Dong H, Sun N, Ran X, et al. All-inorganic Sn-based perovskite solar cells: Status, challenges, and perspectives. ChemSusChem. 2020;**13**(24):6477-6497. DOI: 10.1002/cssc.202001680

[87] Jun Z, Jiajun L, Ximing R, Peijia W, Maxim SM, Yang H, et al. Lead-free perovskite derivative $Cs_2SnCl_{6-x}Br_x$ single crystals for narrowband photodetectors. Advanced Optical Materials. 2019;**7**(10):1900139. DOI: 10.1002/adom.201900139

[88] Hongdan Z, Ludan Z, Jun C, Long C, Chuanqi L, Shuanglong Y. Morphologically controlled synthesis of Cs_2SnCl_6 perovskite crystals and their photoluminescence activity. Crystals. 2019;**9**:258. DOI: 10.3390/cryst9050258

[89] Ullah S, Wang J, Yang P, Liu L, Khan J, Yang SE, et al. Lead-free Cs_2SnI_6 perovskites for optoelectronic applications: Recent developments and perspectives. Solar RRL. 2021;**5**(5): 2000830. DOI: 10.1002/solr.202000830

[90] Dehghan M, Behjat A. Deposition of zinc oxide as an electron transport layer in planar perovskite solar cells by spray and SILAR methods comparable with spin coating. RSC Advances. 2019;**9**: 20917-20924. DOI: 10.1039/C9RA01839E

[91] Penga Y, Cheng Y, Wang C, Zhang C, Xia H, Huang K, et al. Fully doctor-bladed planar heterojunction perovskite solar cells under ambient condition. Organic Electronics. 2018;**58**: 153-158. DOI: 10.1016/j.orgel.2018.04.020

[92] Yue P, Haiyan G, Ziyang H, Jing Z, Yuejin Z. $BiBr_3$ as an additive in $CsPbBr_3$ for carbon-based all-inorganic perovskite solar cells. Journal of Alloys and Compounds. 2020;**835**:155283. DOI: 10.1016/j.jallcom.2020.155283

[93] Haonan S, Qingliang L, Zheng Z, Yong L, Xuhui Y, Guangjie Z, et al. An innovative design of perovskite solar cells with Al_2O_3 inserted at the ZnO/perovskite interface for improving performance and stability. Nano Energy. 2016;**22**:223-231. DOI: 10.1016/j.nanoen.2016.02.025

[94] Deng W, Wenjing L, Weihai S, Xuping L, Guodong L, Zibao W, et al. Guanidinium iodide modification enabled highly efficient and stable all-inorganic $CsPbBr_3$ perovskite solar cells. Electrochimica Acta. 2021;**365**:137360. DOI: 10.1016/j.electacta.2020.137360

[95] Yichuan L, Yu T, Xi W, Jamie CW, Javon MK, Fernando PO, et al. Enhanced optical and electrical properties of polymer-assisted all-inorganic perovskite for light emitting diodes. Advanced Materials. 2016;**28**(40): 8983-8989. DOI: 10.1002/adma.201602513

[96] Zhang W, Saliba M, Stranks SD, Sun Y, Wiesner U, Snaith HJ. Enhancement of perovskite-based solar

cells employing core-shell metal nanoparticles. Nano Letters. 2013;**13**(9): 4505-4510. DOI: 10.1021/nl4024287

[97] Liang PW, Liao CY, Chugh CC, Zuo F, Williams ST, Xin XK, et al. Additive enhanced crystallization of solution-processed perovskite for highly efficient planar-heterojunction solar cells. Advanced Materials. 2014;**26**(22): 3748-3754. DOI: 10.1002/adma.2014 00231

[98] Chauntian Z, Liming D. An 80.11% FF record was achieved for perovskite solar cells by using the NH_4Cl additive. Nanoscale. 2014;**6**:9934. DOI: 10.1039/ C4NR02425G

[99] Chauhan A, Shrivastav AK, Oudhia A. Optical, morphological and electrical studies of fully doctor-bladed $CsPbBr_2Cl$-based perovskite thin film. Microelectronic Engineering. 2022;**258**: 111757. DOI: 10.1016/j.mee.2022.111757

[100] Runzhong L, Ke X. Solvent engineering for perovskite solar cells: A review. Micro and Nano Letters. 2020; **15**(6):349-353. DOI: 10.1049/ mnl.2019.0735

[101] Rachma DS, Veinardi S. Isolation of chlorophyll from spinach and its modification using Fe^{2+} in photostability study. In: Proceedings of the Third International Conference on Mathematics and Natural Science; 23'25 November 2010; Bandung, Indonesia. Bandung, Indonesia: Faculty of Mathematics and Natural Sciences, School of Life Sciences and Technology, and School of Pharmacy at Institute Technologist Bandung; 2010. pp. 859-873

[102] Fan Z, Jun S, Ming C, Junchen L, Yuying H, Yuncai W, et al. Enhanced perovskite morphology and crystallinity for high-performance perovskite solar cells using a porous hole transport layer from polystyrene nanospheres. Physical Chemistry Chemical Physics. 2016;**18**: 32903-32909. DOI: 10.1039/ C6CP06405A

[103] Gupta M, Oudhia A, Bisen DP. A review on perovskite solar cell for high-efficiency future applications. International Journal of Luminescence and Applications. 2017;7:413-416

Chapter 6

Synthesis and Characterization of CZTS Thin Films for Solar Cell Applications

Sadhu K. Kolekar, Rahul Ahir, Kiran Diwate and Haribhahu Borate

Abstract

This chapter provides a comprehensive overview of various deposition techniques used for synthesizing copper zinc tin sulfide (CZTS) thin films, focusing on both vacuum and non-vacuum processes. Vacuum-based methods, known for producing high-quality films with minimal defects, are compared with non-vacuum techniques, which often introduce impurities and defects. The discussion includes techniques, such as sputtering, molecular beam epitaxy (MBE), pulsed laser deposition (PLD), electrodeposition, spray pyrolysis, the sol-gel method, and electron beam evaporation. Among these, spray pyrolysis is highlighted for its scalability, cost-effectiveness, and ability to deposit films at room temperature. The chapter also explores advanced characterization techniques for CZTS thin films, including Kelvin probe force microscopy (KPFM), scanning tunneling microscopy (STM), and X-ray photoemission spectroscopy (XPS). In addition, the chapter discusses the fabrication and optimization of CZTS-based solar cells, emphasizing the impact of pre-treatment, sulfurization temperature, and device architecture on performance. The chapter concludes by addressing the future potential of CZTS thin films in photovoltaic applications.

Keywords: CZTS thin films, solar cells, vacuum and non-vacuum synthesis techniques, synthesis methods, absorber layer, optical absorption coefficient, conversion efficiency

1. Introduction

1.1 CZTS properties, structure, phases

The pursuit of scalable, sustainable, and cost-effective solar energy solutions necessitates the development of stable, earth-abundant, non-toxic, and efficient absorber materials. Kesterite copper zinc tin chalcogenides, specifically Cu_2ZnSnS_4 (CZTS) (copper zinc tin sulfide) and $Cu_2ZnSnSe_4$ (CZTSe) (copper zinc tin sulfide-selenide), have garnered significant attention as promising candidates for the production of low-cost, environmentally friendly thin-film photovoltaics (PV). CZTS, in particular, stands out due to its optimal bandgap of 1.5 eV and high optical absorption coefficient of 10^4 cm^{-1} in the visible spectrum, which together support a theoretical power conversion efficiency (PCE) exceeding 30% [1]. Additionally, the bandgap of CZTS can be readily modified through the incorporation of dopant elements, such as Cd, Se, Ge, Mg, Ni, Co, or Ag, offering flexibility in optimizing material properties for specific applications [2]. Importantly, the quaternary nature of CZTS and CZTSe, consisting of earth-abundant and non-toxic elements, makes these materials well suited for large-scale, eco-friendly production, positioning them as viable alternatives to existing thin-film solar cell absorbers like cadmium telluride (CdTe) and copper indium gallium selenide ($Cu(In,Ga)Se_2$ (CIGS [3, 4]). Beyond their application in photovoltaics, the unique physicochemical and electronic properties of CZTS also make them promising candidates for heterogeneous catalysis, including electrocatalytic hydrogen evolution reactions (HERs) and the photocatalytic degradation of pollutants [5]. Notably, copper-zinc oxide (CZO) heterostructures derived from these materials demonstrate superior performance in hydrogen evolution under sunlight compared to individual copper oxide (CuO) and zinc oxide (ZnO) components [6].

Copper zinc tin sulfide-selenide (CZTSSe) solar cells, despite their favorable optoelectronic properties, have seen limited improvement in efficiency, with the highest recorded efficiency for CZTSSe-based devices stagnating at approximately 12.6% [7]. This stagnation is largely attributed to the presence of antisite disorder, particularly between Cu and Zn atoms, which disrupts the crystal structure. In CZTSSe compounds, Cu, Zn, and Sn cations share similar tetrahedral coordination and comparable ionic radii (Cu + 0.91 Å, Zn2+ 0.88 Å, Sn4+ 0.83 Å), leading to a high likelihood of antisite defect formation [8]. These defects induce potential fluctuations and bandgap tailing, which limit the open-circuit voltage (V_{oc}) of the solar cells. Additionally, other antisite defects, such as Cu-Sn and Zn-Sn, can introduce deep-level defect states, exacerbating charge carrier recombination [9]. The ongoing research and development of kesterite-based (CZTSSe) thin-film solar cells have been driven by their potential to provide a cost-effective, sustainable energy solution. Current efficiencies have reached around 14.9% for CZTSSe absorbers [10], with a theoretical potential efficiency of up to 33% at the radiative limit. However, this is still significantly lower than the 23.6% record efficiency achieved by CIGS solar cells [11]. The performance of CZTSSe solar cells is heavily influenced by the quality of their interfaces, which affects charge carrier transport and recombination rates (**Figure 1**) [13].

Crystallographically, CZTS exists primarily in two forms: the stannite and kesterite structures. These structures differ primarily in the arrangement of Cu and Zn atoms, with the kesterite structure being thermodynamically more stable [14]. CZTS has an optimal direct bandgap between 1.4 and 1.5 eV, along with a high absorption coefficient exceeding 10^4 cm^{-1} in the visible light spectrum [15]. **Figure 2** illustrates the electronic band structure and density of states (DOS) for CZTS, highlighting the energy positions of cation d-orbitals in related cubic semiconductors relative to the valence band maximum (VBM).

The calculated band structure (**Figure 2a**) reveals that CZTS is a direct bandgap semiconductor at the gamma point of the Brillouin zone. The predicted bandgap is approximately 1.98 eV, which aligns well with experimental values reported in the range of 1.78–2.1 eV [12] and previous density functional theory (DFT)-based studies [16]. From this band structure, the effective masses of electrons (m_e^*) and holes

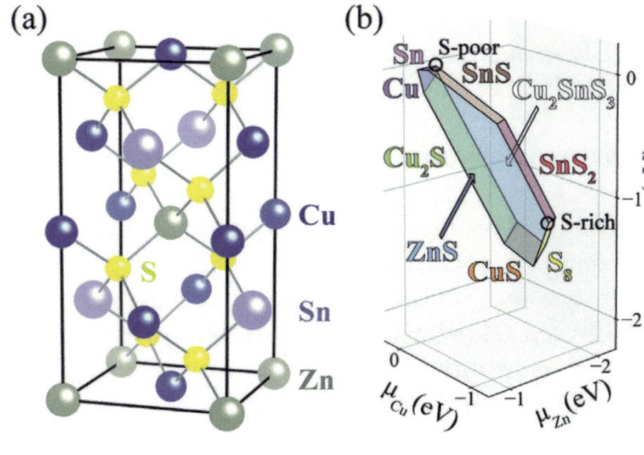

Figure 1.
(a) Atomic structure and (b) phase diagram of CZTS in chemical potential space, where $\mu_i = 0$ represents element i in its standard state. Blue, gray, purple, and yellow balls represent Cu, Zn, Sn, and S atoms, respectively. Reproduced from Ref. [12].

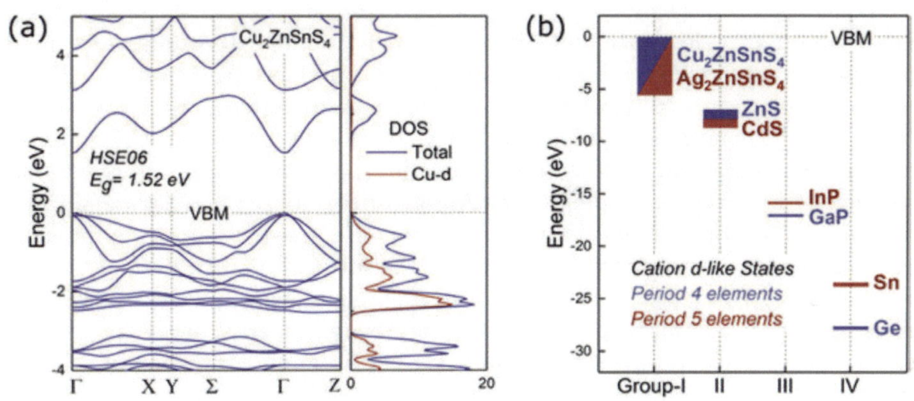

Figure 2.
(a) CZTS electronic band structure and density of states (DOS in units of (eV x primitive cell)$^{-1}$). (b) The energy positions of cation d-orbital projected states in related cubic-like semiconductors. Energies refer to the valence band maximum (VBM). Copyright reproduced from Ref. [16].

(m_h^*) were determined by fitting the conduction band minimum and the valence band maximum, respectively.

This chapter focuses on the deposition and characterization of CZTS (Cu2ZnSnS4) thin films for solar cell applications. In the first section, we present various deposition techniques that facilitate efficient growth of high-quality thin films. The following section covers key characterization methods, including Kelvin Probe Force Microscopy (KPFM), X-ray Photoemission Spectroscopy (XPS), and Ultraviolet Photoemission Spectroscopy (UPS). These techniques were selected for their ability to provide in-depth surface analysis, revealing properties, such as morphology, chemical composition, energy levels, and electrical characteristics of the devices. Advanced characterization techniques offer valuable insights into the material properties and behavior of CZTS in solar cells. KPFM is employed to study surface potential, electronic work function, and charge distribution, offering nanometer-scale resolution. XPS provides detailed information about the composition of surfaces, chemical reactions, and electronic structures at interfaces. The chapter concludes with a summary of findings and discusses future research directions for further optimizing CZTS-based solar cells.

2. Thin-film deposition methods

Thin-film deposition involves the process of applying thin coatings onto a substrate material, playing a pivotal role in the development of solar cell devices. The deposition of high-quality thin films is essential to enhance the efficiency and performance of CZTS-based solar cells. These thin coatings exhibit various properties that can be leveraged to optimize the performance of devices. Although the term "thin" is relative, in thin-film deposition techniques, it typically refers to layers just a few tens of nanometers thick. While thin films are commonly associated with semiconductors, their significance extends across multiple fields. In solar cell technology, thin films are crucial as they modify the surface interactions of the device, differentiating the properties of the film from those of the bulk substrate. This distinction is particularly important for optimizing energy conversion in solar cells. CZTS (Cu_2ZnSnS_4) has emerged as a promising material for solar cell applications due to its abundance, low cost, and environmentally friendly, non-toxic components. As an absorber layer, CZTS exhibits favorable optical properties, including a direct bandgap energy of approximately 1.5 eV and a high absorption coefficient of around 10^4 cm^{-1} [17]. Various deposition methods have been used to fabricate CZTS films, including sputtering [18], electro-deposition [19], sol-gel methods [20], pulsed laser deposition [21], and spray pyrolysis [22]. Among these, spray pyrolysis stands out as a non-vacuum technique that is both versatile and effective for depositing films over large areas. Several studies have demonstrated the successful fabrication of CZTSSe ($Cu_2ZnSn(S,Se)_4$)-based solar cells using this method. However, many of these devices employed selenium-rich compositions, raising concerns due to the toxicity of selenium. Consequently, pure sulfide-based CZTS, which avoids the use of selenium, is preferred for solar cell applications. Further research has shown that annealing CZTS thin films with sulfurization on indium-doped tin oxide-coated glass substrates enhances the performance of solar cells [23].

In the spray pyrolysis method, a precursor solution is sprayed onto a preheated substrate, where the components react to form the desired chemical compound. This non-vacuum technique offers several advantages, including the following features: (i) it eliminates the need for high-quality targets and substrates, (ii) stoichiometry of the films can be easily controlled by adjusting the precursor concentrations, (iii) doping can be achieved by adding a soluble salt of the desired dopant to the solution, (iv) there are no restrictions in the size or surface profile of the substrate, (v) large-area deposition is possible, and (vi) vacuum processes are not required, making this technique highly beneficial for industrial applications [24]. For instance, an efficiency of 8.1% was achieved for a CZTS thin film with Cu-poor and Zn-rich composition. This film was fabricated using spray pyrolysis of an aqueous solution containing copper II nitrate (Cu(NO3)2), zinc nitrate (Zn(NO3)2), tin(II) methanesulfonate (Sn(CH3SO3)2), and thiourea, followed by sulfurization at 600°C for 30 minutes on a molybdenum (Mo)-coated glass substrate [25]. In many sol-gel approaches, solvents like 2-methoxyethanol are used, which are expensive, toxic, and require low-temperature processing [26]. To address these concerns, dimethyl sulfoxide (DMSO) has been proposed as a low-cost, non-toxic alternative. DMSO has been proven to dissolve various metal salts and thiourea effectively, enabling good control over the metal composition of the precursor solution, which results in the desired absorber composition [27]. Thus, the use of DMSO presents a promising route for developing low-cost and environmentally friendly solar cells.

One of the primary challenges in producing CZTS solar cells lies in obtaining a single-phase CZTS film. The CZTS phase diagram indicates that the stoichiometric single-phase region is quite narrow, making it difficult to synthesize impurity-free, stoichiometric CZTS films. Therefore, various parameters, including the atomic ratio of the solution, carrier gas pressure, spray rate, and movement speed of the spray head, must be carefully controlled during the deposition process. Among these factors, substrate temperature plays a critical role in reducing the formation of secondary phases and improving the overall film quality. It has been observed that compositional ratios of Cu/(Zn + Sn) between 0.8 and 0.9 and Zn/Sn between 1.2 and 1.3 are often associated with improved physical properties of CZTS films and higher solar cell efficiencies [28]. From an economic standpoint, non-vacuum techniques are preferable due to their lower costs. These techniques, including spray pyrolysis, are more affordable and versatile for CZTS thin-film deposition, though they may introduce impurities and result in multiple coatings or cracks. This study explores the synthesis of stoichiometric CZTS films using a non-toxic, DMSO-based sol-gel solution. The films are deposited via spray pyrolysis at substrate temperatures of 250, 300, and 350°C, without requiring sulfurization or toxic atmospheres like hydrogen sulfide (H2S) or sulfur vapor [23]. Structural, compositional, morphological, and optical properties of the films are investigated, with the proposed method offering a promising, cost-effective, and environmentally friendly approach to CZTS film fabrication, free of impurities and structural defects.

In the production of thin films for solar cell applications, several deposition techniques are often combined to optimize material properties. Both physical and chemical deposition methods are utilized. Physical techniques include Ionized Cluster Beam (ICB) deposition, Electron Cyclotron Resonance (ECR) plasma, and reactive sputtering. On the other hand, chemical processes, such as Chemical Vapor Deposition (CVD)—encompassing variations like ECR-plasma-CVD, sol-gel processing, Sequential Ionic Layer Adsorption and Reaction (SILAR), chemical spray pyrolysis, hot-wire CVD, and electron beam evaporation—are commonly applied.

2.1 Vacuum-based deposition approaches for CZTS thin films

For solar cell applications, achieving high-quality thin films with uniformity, minimal defects, and strong crystallinity is essential. These characteristics can be effectively realized through vacuum-based deposition methods, which offer significant advantages for fabricating high-performance CZTS thin-film devices. In these approaches, the constituent elements of CZTS are deposited onto a substrate via sputtering or evaporation/co-evaporation processes under optimized conditions of pressure and temperature. These methods provide precise control over the chemical composition of the thin films and are known for their reproducibility, making them suitable for large-scale production.

Vacuum-based techniques can be categorized into several methods, including sputtering, evaporation, pulsed laser deposition (PLD), and molecular beam epitaxy (MBE). Each of these techniques has been demonstrated to contribute to the production of high-quality CZTS films for photovoltaic applications due to their ability to manipulate film composition and structure effectively.

2.2 Non-vacuum-based approaches for CZTS thin-film fabrication

In recent years, non-vacuum-based methods have gained significant attention for the fabrication of CZTS thin films used in photovoltaic (PV) devices. While vacuum-based techniques were traditionally the dominant approach, they are often associated with high costs, low material utilization, high-energy consumption, and challenges in scaling up production. In contrast, non-vacuum deposition methods offer a more cost-effective alternative, with reduced energy demands and simpler scalability. These methods produce uniform and high-quality thin films, making them attractive for large-scale production.

The advantages of non-vacuum techniques, such as simplicity, high throughput, and minimal process complexity, have contributed to their increasing adoption for CZTS thin-film production. The availability of abundant and low-cost materials used in these processes further drives the growth of CZTS-based solar cell manufacture. The shift toward non-vacuum methods marks a significant development in the field, particularly as the industry seeks to lower production costs and enhance efficiency. Several studies have demonstrated the effectiveness of non-vacuum approaches in producing CZTS thin films with desirable photovoltaic properties.

2.2.1 The spray pyrolysis deposition (SPD) technique

The spray pyrolysis deposition (SPD) technique is widely adopted for fabricating thin films due to its simplicity, ease of handling, and suitability for large-area deposition, particularly in device fabrication applications. The scalability of this method, combined with its straightforward design and reliable reproducibility, makes it highly attractive for industrial use. One of the significant advantages of SPD is its ability to operate without the need for a vacuum environment at any stage of deposition, which further enhances its industrial appeal.

Numerous studies have explored the potential of SPD for producing CZTS thin films, with Nakayama and Ito (1996) being the first to report successful fabrication using this method [29]. In their work, a precursor solution containing cuprous chloride (CuCl), zinc chloride ($ZnCl_2$), stannic chloride ($SnCl_4$), and thiourea dissolved in a 50:50 mixture of distilled water and ethanol was sprayed onto single-layer graphene (SLG) substrates at temperatures ranging from 280–360°C. The resulting stoichiometric films, with a stannite crystal structure oriented along the (112) plane, exhibited a bandgap energy of 1.46 eV after annealing at 550°C in an argon (Ar)/ H_2S atmosphere [30]. Further advancements in CZTS thin-film synthesis via SPD were reported by Kamoun et al. [31], who optimized deposition conditions through two experimental series with spray durations of 30 and 60 minutes. Their X-ray diffraction (XRD) analysis revealed peaks corresponding to the (112), (200), (220), and (312) planes of kesterite-structured CZTS. However, secondary phases, such as cuprous sulfide (CuS) and cupric sulfide (Cu_2S), were detected at higher substrate temperatures. The best crystallinity was achieved for films deposited at 340°C for 60 minutes, with a bandgap energy of 1.5 eV [29]. Kumar et al. [32] optimized the fabrication process of CZTS thin films to achieve a single phase without the need for annealing or post-deposition sulfurization. In this method, a precursor solution composed of cupric chloride (0.01 M), zinc acetate (0.005 M), stannic chloride (0.005 M), and thiourea (0.04 M) was sprayed onto heated soda-lime glass (SLG) substrates at temperatures ranging from 290–450°C. XRD analysis revealed the presence of Cu2SnS3 and copper sulfide (CuxS) phases, along with CZTS formation

DOI: http://dx.doi.org/10.5772/intechopen.1008712

at 290°C, while zinc sulfide (ZnS) appeared at 450°C. The films were mostly copper- and zinc-rich but sulfur-deficient, with optical bandgaps ranging between 1.40 and 1.45 eV. Rajeshmon et al. [33] also explored CZTS thin films for solar cell applications, employing the spray pyrolysis technique. The films were deposited using copper chloride and zinc chloride as copper and zinc sources, respectively, while stannous chloride and stannic chloride were used as tin precursors. Thiourea served as the sulfur source. The precursor solution, consisting of cuprous chloride (0.02 M), zinc acetate (0.0125 M), stannous/stannic chloride (0.0125 M), and thiourea (0.15 M), was sprayed onto SLG substrates at 350°C at a rate of 8 mL/min. To form p-n junctions, CZTS films were deposited on tin-doped indium oxide (ITO) glass substrates, followed by an indium sulfide (In2S3) buffer layer deposition at 350°C via chemical spray pyrolysis. Solar cells prepared with stannous chloride showed promising junction properties, although no efficiency data were provided. On the other hand, cells prepared with stannic chloride demonstrated an open-circuit voltage (V_{oc}) of 380 mV and a short-circuit current density (J_{Sc}) of 2.4 mA/cm^2. The stannic chloride precursor produced CZTS films with properties suitable for solar cell absorber layers (**Figure 3**).

2.2.1.1 Electro-deposition technique

Electro-deposition is a widely utilized technique for fabricating semiconductor thin films. It is suitable for both small-scale research and large-scale industrial applications due to its simplicity, cost-effectiveness, and scalability. This method has been commonly used for fabricating absorber layers in solar cells, such as cadmium telluride (CdTe) and copper indium gallium selenide (CIGS) [35]. Recently, its application has been extended to the fabrication of copper zinc tin sulfide (CZTS) thin films.

There are two primary approaches for preparing CZTS thin films through electro-deposition: (I) sequential electroplating of precursor materials and (II) single-step electro-deposition of the entire precursor. These films are then subjected to sulfurization or annealing in various atmospheres, such as sulfur vapor (S), sulfur and nitrogen (S2 + N2), or sulfur and argon (S2 + Ar), at temperatures ranging from 300–600°C. Scragg et al. [36] were among the first to attempt CZTS thin-film fabrication

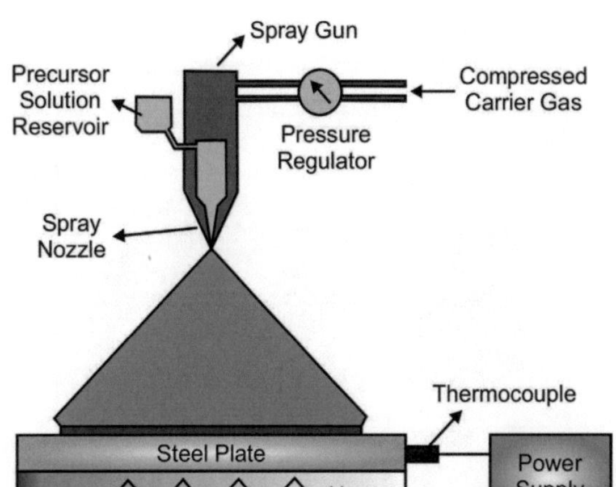

Figure 3.
Schematic of spray pyrolysis depositions (Ref. [34]).

using electro-deposition. In their process, a sequence of electro-deposition of copper (Cu), zinc (Zn), and tin (Sn) was performed on molybdenum (Mo)-coated soda-lime glass (SLG) substrates, followed by annealing in a sulfur atmosphere at 550°C for 2 hours. The X-ray diffraction (XRD) analysis revealed the formation of CZTS, though secondary phases like tin sulfide (SnS2) were also detected (**Figure 4**).

The solar cells are fabricated from these films and exhibited a power conversion efficiency (PCE) of 0.8%, with an open-circuit voltage (V_{oc}) of 295 mV, a short-circuit current density (J_{Sc}) of 8.7 mA/cm^2, and a fill factor (FF) of 0.32. Similarly, Araki et al. [37] explored the electro-deposition method for CZTS fabrication by sequentially depositing stacked layers of Cu, Sn, and Zn onto Mo-coated glass substrates. Copper sulfate (CuSO4), tin sulfate (SnSO4), and zinc sulfate (ZnSO4) were used as the electrolyte sources for the respective metal layers. The stacked metal precursors were then sulfurized in a sulfur and nitrogen (S2 + N2) atmosphere at temperatures between 300 and 600°C. XRD analysis showed distinct CZTS peaks for films processed at temperatures above 400°C. However, films processed at 600°C exhibited Zn-rich, Sn-rich, and sulfur-poor compositions.

These early studies on the electro-deposition of CZTS thin films laid the groundwork for further improvements in the efficiency and composition control of CZTS-based solar cells.

2.2.2 Sol-gel-based spin coating technique for CZTS thin-film fabrication

The sol-gel spin coating method is a well-established approach for the fabrication of semiconductor thin films. This process primarily consists of two major steps: (1) preparing a sol-gel solution from precursors and (2) spin coating the solution onto a substrate to achieve the desired thin-film structure. This direct liquid deposition method, which includes solution, particle, and mixed particle-solution precursors, is particularly advantageous for large-area manufacture, making it highly relevant in industrial applications.

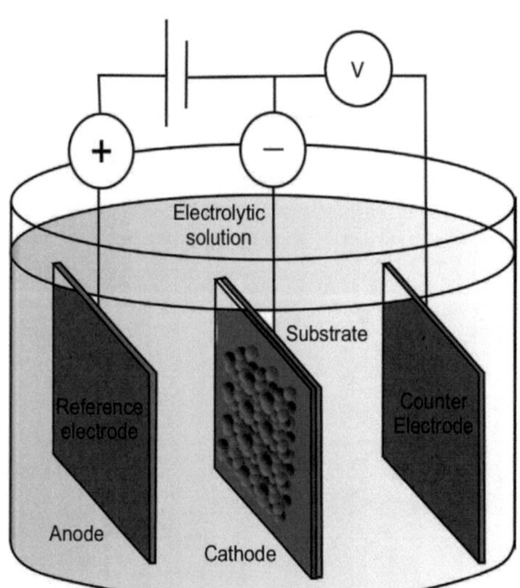

Figure 4.
Schematic of electro-deposition setup.

In the context of CZTS (Cu_2ZnSnS_4) thin-film production, typical precursors used include copper acetate monohydrate, zinc acetate dihydrate, tin chloride dihydrate, 2-methoxyethanol, and monoethanolamine. Tanaka et al. [38] pioneered the application of the sol-gel technique to fabricate CZTS thin films. In their method, the precursor solution was stirred at 45°C for 1 hour, followed by spin coating at 4000 rpm (revolutions per minute) for 30 seconds. The precursor films were dried at 300°C, with multiple iterations of the coating and drying process to achieve the desired thickness. Subsequently, the films were annealed in an N_2 + H_2S atmosphere at 500°C for 1 hour. X-ray diffraction (XRD) analysis confirmed the formation of single-phase CZTS without any secondary phases, such as oxides or hydroxides. The resulting films exhibited a near-stoichiometric composition with a bandgap energy of 1.49 eV. Following this breakthrough, Tanaka et al. [39] reported the fabrication of CZTS-based solar cells using a fully non-vacuum process to reduce production costs. In this case, the sol-gel sulfurization technique was employed, where the sol solution was stirred at 60°C for 1 hour and spin-coated onto soda-lime glass (SLG) substrates at 3000 rpm for 30 seconds. XRD patterns revealed peaks corresponding to CZTS, molybdenum (Mo), and molybdenum disulfide (MoS_2). These films were found to be Cu-poor and Zn-rich. The resulting solar cell, configured as SLG/Mo/CZTS/CdS/Al/Al, demonstrated a conversion efficiency of 1.01% with an open-circuit voltage (V_{OC}) of 390 mV, a short-circuit current density (J_{Sc}) of 7.8 mA/cm^2, and a fill factor (FF) of 0.33. The relatively low efficiency was attributed to the higher resistivity of sol-gel spin-coated Al compared to sputtered Al. This marked the first report of a solar cell, in which all semiconductor layers were prepared via non-vacuum processes.

In subsequent work, CZTS thin films were prepared under non-vacuum conditions on two different substrates: Mo-coated SLG and bare SLG. The sol-gel solution was spin-coated at 3000 rpm for 30 seconds, followed by annealing in an N_2 + H_2S atmosphere at 500°C for 1 hour. A solar cell with a structure of Al/Al/CdS/CZTS/Mo/SLG, fabricated using a CZTS thin film with a Cu/Zn + Sn ratio of 0.80, exhibited an improved efficiency of 2.03% (V_{oc} = 575 mV, J_{Sc} = 9.69 mA/cm^2, and FF = 36.4%) [40]. This represents the highest reported efficiency for pure CZTS solar cells fabricated using the sol-gel technique.

2.2.3 Chemical vapor deposition (CVD) in thin-film fabrication

Chemical Vapor Deposition (CVD) is a widely adopted method for producing thin films, bulk materials, powders, and composites. In CVD, gaseous reactants undergo a chemical reaction on or near a heated substrate to form a thin film, as depicted in **Figure 5**. The process typically begins with volatile precursors flowing into a chamber containing the heated substrate. These precursors decompose or react upon reaching the substrate surface, forming the desired thin film. Any unreacted precursors and by-products are then exhausted from the chamber. One of the primary advantages of CVD is its ability to produce conformal films, where the film thickness on sidewalls is comparable to that on the top surface. This makes CVD particularly suitable for coating complex, elaborately shaped pieces, including both inner and outer features. It is also highly effective at filling high-aspect-ratio holes. Moreover, CVD can produce films with relatively high purity, as impurities in volatile precursors can be removed through distillation.

Compared to physical vapor deposition (PVD) techniques, CVD does not require as high a vacuum and generally provides higher deposition rates. However, CVD has certain limitations. It requires volatile precursors at near-room temperatures,

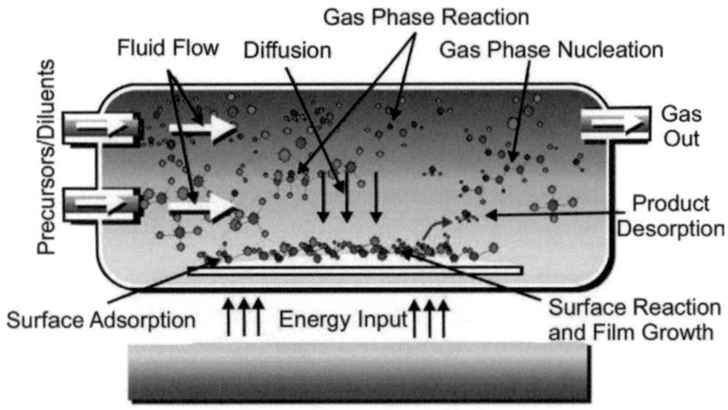

Figure 5.
Schematic diagram for chemical vapor deposition (CVD). (https://www.mks.com/n/cvd-physics).

some of which may be hazardous. For instance, nickel carbonyl (Ni(CO)4) is highly toxic, diborane (B2H6) is explosive, and silicon tetrachloride (SiCl4) is corrosive. Furthermore, the by-products of CVD reactions, such as carbon monoxide (CO), hydrogen (H_2), and hydrogen fluoride (HF), can pose significant risks. Another challenge of CVD is its requirement for elevated deposition temperatures, which can limit the types of substrates used due to thermal stresses and mechanical instabilities in the deposited films. The selection of the appropriate CVD process and reactor type is influenced by factors, such as the substrate material, film thickness, uniformity, precursor cost, and desired coating morphology. To enhance deposition rates or reduce operating temperatures, various modified CVD processes incorporate plasma, ions, lasers, or combustion reactions.

2.2.4 Sputtering systems in thin-film deposition

Sputtering is a physical vapor deposition (PVD) technique where energetic particles, such as accelerated ions, collide with a target material, causing the ejection of surface atoms. These atoms travel through the vacuum chamber and deposit onto the substrate, forming a thin film. Various sputtering methods are employed for thin-film deposition, each tailored to specific materials and applications.

One of the simplest sputtering systems is the direct current (DC) diode sputtering system, which consists of two planar electrodes: an anode and a cold cathode. The target material, placed on the front surface of the cathode, is sputtered by argon ions produced during a glow discharge. The anode, which holds the substrate, is positioned opposite the cathode. The sputtering gas, typically argon, is introduced into the chamber at a pressure of about 0.1 Torr. When a DC voltage is applied between the electrodes, a glow discharge is sustained, leading to the deposition of thin films on the substrate. This system works efficiently with metal targets, as metallic electrodes help maintain the glow discharge [41]. However, in DC sputtering systems, using an insulating material as the target poses challenges. The build-up of a surface charge from positive ions on the insulator's surface interrupts the discharge. This issue is overcome by using radio-frequency (RF) diode sputtering, where an RF voltage is applied to the insulator target, allowing continuous discharge and effective deposition of insulating thin films [42]. Reactive sputtering is employed when thin-film compounds are required. This process involves the introduction of reactive gases, such as hydrogen (H_2), oxygen (O_2), or nitrogen (N_2), into the chamber. Reactive sputtering can be applied to both DC and RF diode systems, enabling the deposition of compound films directly from the target material.

Despite extensive research on CZTS heterojunction devices, the power conversion efficiency (PCE) of CZTS solar cells, currently at 11.4%, remains significantly lower than those of competing technologies like CIGS and CdTe solar cells [43]. Several factors contribute to this performance gap, with the complex growth mechanism of CZTS layers and their compositional dependence being central challenges. The choice of molybdenum (Mo) as a back contact material significantly affects the fundamental properties of CZTS absorber layers. Mo and CZTS exhibit a substantial lattice mismatch, with their respective lattice constants being 5.43 and 3.15 Å, and thermal expansion coefficients of 12 and 7 ppm/°C, respectively [44]. This mismatch induces strain at the Mo/CZTS interface, which often results in high dislocation densities and defect formation, complicating the fabrication process and potentially degrading device performance [45]. CZTS has a narrow phase stability range, leading to structural inhomogeneity and non-stoichiometry, which further complicates its growth. The strain introduced in the films during deposition and annealing can be detrimental, causing structural disorder and affecting the material's electrical and optoelectronic properties. One of the key challenges is Cu-Zn disorder, which plays a pivotal role in determining the material's properties [45]. The order parameter (S), which describes the degree of order in the CZTS crystal, is highly sensitive to the heat treatment history of the films [46]. Considerable research has focused on optimizing the heat treatment process, especially the sulfurization step, to improve CZTS film quality. Sulfur partial pressure and sulfurization time during annealing are critical parameters that strongly influence the structural and optoelectronic properties of CZTS [47]. Proper control of these parameters is essential for achieving ordered films and minimizing Cu-Zn disorder.

Doping and alloying in CZTS have emerged as effective strategies to improve efficiency by tuning optoelectronic properties and reducing inherent disorder in the crystal structure. However, controlling key parameters, such as sulfur partial pressure and doping levels, remains challenging, leading to fabrication complexities. These difficulties underscore the need for new methodologies and systematic approaches to reduce structural disorder while avoiding excessive strain in CZTS films. By addressing these challenges, it is possible to push the limits of CZTS solar cell efficiency and develop more reliable fabrication techniques that can yield high-performance devices.

The incorporation of a high-temperature seed layer (SL) plays a critical role in enhancing the growth of CZTS thin films. This approach improves film quality by enhancing adhesion to the substrate and optimizing both the optoelectronic and structural properties of the material. The introduction of SL growth significantly increases crystallinity, reduces strain, and enhances the crystal grain size from 86 to 138 nm. Moreover, strain modulation leads to improved Cu-Zn ordering, which is confirmed by a decrease in Urbach energy (EU) and an increased bandgap in SL-grown samples. The improved ordering directly influences the optoelectronic performance, notably enhancing hole mobility from 2.13 to 13.5 $cm^2/V{\cdot}s$. [44]. The ordering induced by the seed layer also affects the photoluminescence (PL) spectrum, revealing a dominant band-to-band (BB) transition in SL-based CZTS films. Raman spectroscopy further supports this, showing a shift toward higher Raman frequencies, which is attributed to strain modulation through SL growth.

The effect of the SL growth on device performance was evaluated by fabricating a metal-semiconductor-metal (MSM) photodetector (PD). The SL-assisted CZTS photodetector demonstrated faster charge carrier transport compared to its counterpart without the SL, a direct result of the improved material properties. The improved structural and optoelectronic characteristics attributed to the SL are shown in **Figures 6(a)–(c)**, depicting the fabrication process and thermal profile. Microscopic analysis of the films revealed that CZTS films grown without SL showed smaller grains with random grain size distribution. In contrast, the SL-CZTS samples exhibited larger and more compact grains with minimal grain size variation. Energy-dispersive X-ray spectroscopy (EDS) showed a Cu-rich and Zn-poor composition in the SL samples, which leads to a higher degree of Cu-Zn disorder. However, the exact impact of the seed layer on the stoichiometry is not yet fully understood. Adhesion to the Mo/SLG substrate was found to be better in SL-CZTS films, whereas the without seed layer (W/O SL) samples exhibited small pinholes. X-ray diffraction (XRD) results confirmed the CZTS phase as dominant, with the highest intensity peak corresponding to the (112) orientation, followed by (200), (220), and (312) planes. Notably, the W/O SL samples displayed secondary phases, which were absent in the SL-grown films. To further assess the effect of SL growth on crystallinity and strain, the Williamson-Hall (W-H) method was applied. This method is widely used to analyze strain-induced broadening and to estimate crystallite size in polycrystalline films.

2.2.5 Pulsed laser deposition (PLD)

Pulsed Laser Deposition (PLD) has emerged as a versatile and powerful technique for thin-film growth, particularly for complex oxides and quaternary chalcogenides. Since its inception with the ruby laser, PLD has seen significant advancements in both its experimental setup and precision, making it one of the most reliable methods for material deposition in modern research. Initially, the process posed challenges

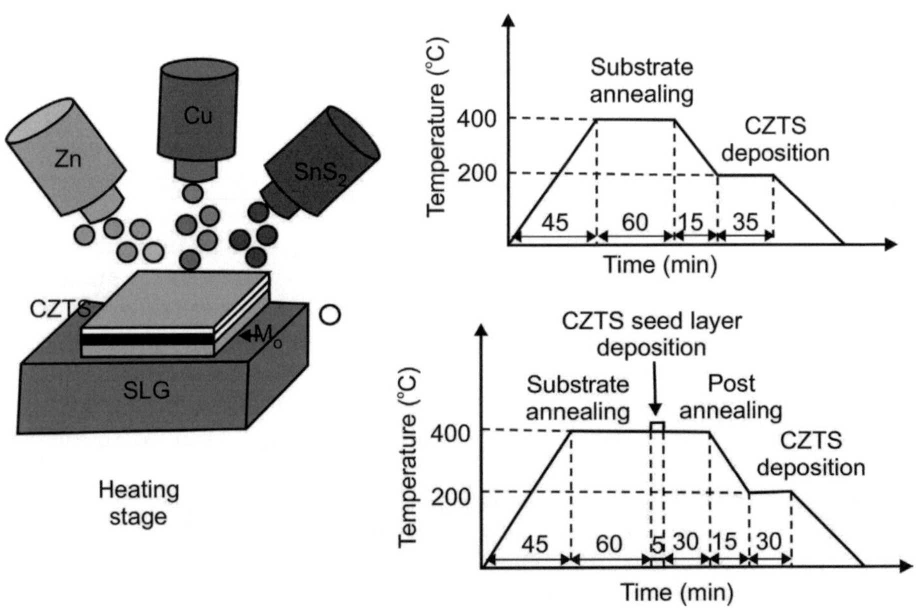

Figure 6.
(a) Illustration of CZTS deposition via co-sputtering of Cu, Zn, and SnS2; temperature profiles for CZTS (b) W/O Seed Layer and (c) with Seed Layer growth, respectively. [reproduced Ref. [44]].

in control, but with advancements in laser technology and vacuum systems, it has evolved into a more refined technique that supports the deposition of high-quality thin films. The PLD process involves four distinct stages: laser-target interaction, ablation and plume formation, deposition on the substrate, and film growth. In this method, a high-energy laser pulse is directed at a target material within a vacuum chamber, causing the material to evaporate and form a plume of particles, which subsequently condense on the substrate surface. Several factors can affect the quality of the resulting thin film, including the composition of the target material, the parameters of the laser (such as energy density and pulse duration), the type and temperature of the substrate, and the pressure and distance within the chamber. Fine-tuning these parameters enables control over film properties, allowing the production of amorphous, polycrystalline, or even epitaxial layers. However, PLD does have limitations, such as the potential for molten particles in the film, which can lower quality if not properly managed. Despite these challenges, its capacity to handle a diverse array of materials and the precise control it offers over film characteristics make PLD a leading choice in thin-film research (**Figure 7**).

The fabrication of CZTS (Cu_2ZnSnS_4) thin films with a well-controlled stoichiometric composition is essential to achieve high-performance solar cells. However, achieving this level of control during the co-evaporation process is challenging due to the volatility of elements, such as Sn, Zn, and S. Controlling the evaporation rate of these elements is complex, particularly during the co-evaporation method, which often results in non-uniform film properties. An alternative technique involves the sputtering of metal precursors followed by a post-sulfurization step. Katagiri et al. [49] demonstrated this approach using radio-frequency (RF) co-sputtering of Cu, ZnS, and SnS targets, followed by sulfurization, achieving an efficiency of 6.77% for the resulting CZTS thin-film solar cell. However, the sulfurization process often results in rough morphologies and non-uniformity in film composition, leading to challenges in scalability and reproducibility.

Pioneering work by Moriya et al. [50, 51] introduced the use of PLD for CZTS solar cells. In their study, the CZTS thin films were annealed at 500°C in an N_2 atmosphere, resulting in Sn-rich films with a bandgap energy of around 1.5 eV. A solar cell based on a CZTS absorber layer was fabricated, using an Al window layer and a CdS buffer layer. The device, with an active area of 0.092 cm^2, achieved an open-circuit

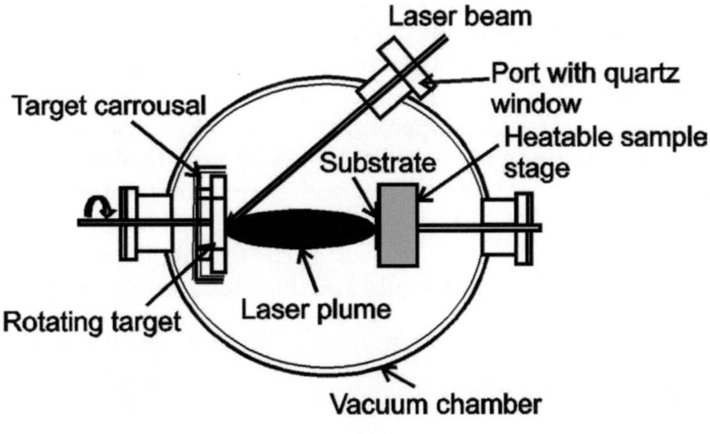

Figure 7.
Schematic diagram of pulsed laser deposition (PLD) (Ref. [48]).

voltage (V_{oc}) of 546 mV, a short-circuit current (J_{sc}) of 6.78 mA/cm^2, a fill factor (FF) of 0.48, and a conversion efficiency of 1.74% [50, 51]. Despite these advances, achieving a stoichiometric composition remains a challenge. Non-stoichiometric compositions, particularly sulfur deficiency, have been a significant issue. To address this, Moriya et al. explored annealing the CZTS precursors in N_2 and N_2 + H_2S (5%) atmospheres. Annealing in the N_2 + H_2S atmosphere successfully suppressed sulfur loss, leading to nearly stoichiometric films. The bandgap of the films annealed at 500°C in the N_2 + H_2S atmosphere remained at 1.5 eV. However, the efficiency of the solar cell fabricated from these films, with an active area of 0.12 cm^2, was relatively low, with a V_{oc} of 336 mV, a J_{sc} of 6.53 mA/cm^2, a FF of 0.46, and a conversion efficiency of 0.64% [51]. The decrease in efficiency, despite stoichiometric composition, was attributed to the film's compositional ratio. Moriya et al. suggested that a Zn-rich and Cu-poor composition is essential for achieving higher efficiency in CZTS thin-film solar cells. Therefore, optimizing the precursor composition remains a critical factor in improving the performance of these devices.

2.2.6 Atomic layer deposition (ALD)

Atomic Layer Deposition (ALD) offers significant advantages in the fabrication of thin films, particularly due to its precise control over material thickness and composition. Unlike other deposition methods, such as Chemical Vapor Deposition (CVD) and Physical Vapor Deposition (PVD), ALD is characterized by its cyclic, self-limiting process, which results in uniform and conformal coatings, even on complex surfaces. This makes ALD highly suitable for modern device architectures, where smaller and more intricate structures are increasingly required. The ALD process operates through sequential pulses of gaseous precursors, which react with the substrate in a series of distinct steps known as "half-reactions." These half-reactions contribute to the controlled synthesis of the thin film [52]. Each precursor is introduced into the reaction chamber under vacuum conditions (typically below 1 Torr), where it undergoes a self-limiting reaction with the surface. This ensures that no more than one monolayer of material is deposited at each step. Following the reaction, an inert carrier gas, usually nitrogen (N_2) or argon (Ar), purges the chamber to remove any unreacted precursors and by-products. The process is then repeated with a counter-reactant, gradually building up the desired film thickness layer by layer.

The deposition temperature plays a crucial role in the ALD process. ALD typically occurs at moderate temperatures, often below 350°C, within what is known as the "ALD temperature window." Operating within this window is essential for achieving high-quality film growth, as temperatures outside this range can lead to undesirable outcomes. At lower temperatures, reaction kinetics may slow down, or precursors may condense on the substrate, hindering growth. Conversely, higher temperatures may cause precursor decomposition or rapid desorption, leading to non-ALD-type deposition. Therefore, maintaining optimal conditions within the ALD temperature window is critical to leverage the technique's advantages, including excellent control over film uniformity and thickness, which are essential for modern device fabrication [52].

2.2.7 Molecular beam epitaxy (MBE)

Molecular Beam Epitaxy (MBE) is a widely used technique for epitaxial growth on crystalline substrates. The process involves directing molecular or atomic beams onto a heated substrate, allowing controlled layer-by-layer growth. The materials used are placed in evaporation cells, where they are heated to form beams with angular dispersion. During deposition, the substrate is continuously rotated and heated to ensure homogeneity. MBE is performed in an ultra-high vacuum environment, typically at pressures of around 10^{-8} Pa, which ensures minimal contamination during the deposition process [53]. A key advantage of MBE is its slow deposition rate, generally below 1000 nm/s, which enables precise control over the growth of epitaxial layers. The term "beam" refers to the fact that evaporated atoms or molecules travel in straight paths and do not interact with one another or with gases in the vacuum chamber, thanks to the long mean free path of the particles. This characteristic is crucial for achieving high-quality epitaxial growth. MBE is also used for organic semiconductors, where entire molecules, rather than individual atoms, are evaporated and deposited onto the substrate [54].

Variants of Molecular Beam Epitaxy (MBE) techniques have evolved to enhance control over film deposition and improve material quality. While traditional MBE involves vapor-phase deposition of monomer flux onto a substrate, several advanced methods refine this process by modulating the energy and characteristics of the deposited beam. One such method is Interrupted (or Pulsed) MBE, where flux modulation controls surface supersaturation, influencing island nucleation rates and leading to smoother films with improved surface morphology. Another variant, Ion-assisted MBE, uses low-energy ion bombardment (30–100 eV) during deposition to affect nucleation, impurity pinning, and vacancy generation, while also allowing subsurface recoil implantation. Ultra-high Vacuum (UHV) Sputtering, which utilizes an energetic flux (10–150 eV), increases surface mobility and enables subsurface implantation, particularly useful for penetrating or modifying hydrogenated surfaces. These advancements allow for precise control over film growth, resulting in high-quality materials with customizable properties.

3. CZTS thin films characterization techniques

Characterizing CZTS thin films is challenging due to various defects in both surface and bulk regions, which are difficult to visualize at micro- and nanoscales. Differentiating between surface and bulk defects is also complex but essential for improving CZTS-based solar cells. Examining factors like surface grain boundaries, chemistry, and physics, along with the influence of defects on surface potential, provides insights to enhance device performance. This section focuses on advanced characterization techniques, including surface potential measurements and topological analysis, which are crucial for understanding CZTS interface properties. While traditional methods like scanning electron microscopy (SEM) and XRD offer morphological and crystallographic information, advanced methods such as X-ray Photoemission Spectroscopy (XPS) are highlighted for their ability to reveal electronic structure and defect information within CZTS films.

X-ray photoelectron spectroscopy (XPS) is an essential analytical tool for investigating the surface properties of materials, particularly in the context of thin films used in solar cell devices. This technique is widely employed for the analysis of semiconductor/electrode junctions, heterojunctions, and passivation layers in solar cells. Nisika et al. [55] investigated the interface properties of CZTS bulk films with cadmium sulfide (CdS) and titanium dioxide (TiO2) buffer layers using X-ray photoelectron spectroscopy (XPS). Their core-level XPS spectra analyses for elements, such as Sn 3d, Zn 2p, Cu 2p, and S 2p, revealed consistent stability across the CZTS-TiO2 and CZTS-CdS interfaces, showing no notable variations (**Figure 8a-d, f-i**). However, shifts in the Ti 2p, Cd 3d, and S 2p spectra indicated alterations in the surrounding chemical environment. Distinct peak separations of 8.4, 23.1, and 19.8 eV were associated with Sn (IV), Zn (II), and Cu (I), respectively. Notably, Sn 3d and Zn 2p peaks were absent at the CZTS-CdS interface, indicating Sn and Zn depletion (**Figure 8f, g**). Sulfur's core levels exhibited doublets at 161.75 and 162.83 eV with a 1.1 eV split (**Figure 8d, i**), while Ti 2p levels, split into 2p3/2 and 2p1/2 states at 458.94 and 464.75 eV with a 5.6 eV separation, confirmed Ti (IV) valence and TiO2-phase formation.

Ultraviolet photoelectron spectroscopy (UPS) is a valuable analytical tool in materials science, offering surface-sensitive insights into electronic structure, work functions, and energy levels across various materials. Widely applied to semiconductor and photovoltaic research, UPS provides direct measurements of the work function in heterojunctions, making it instrumental in studying interface layers, band edge alignment, and device-relevant states, such as defect or interface states. This technique's ability to probe only the top few nanometers of a material enables precise assessments of surface chemistry and the influence of surface treatments on the overall device behavior.

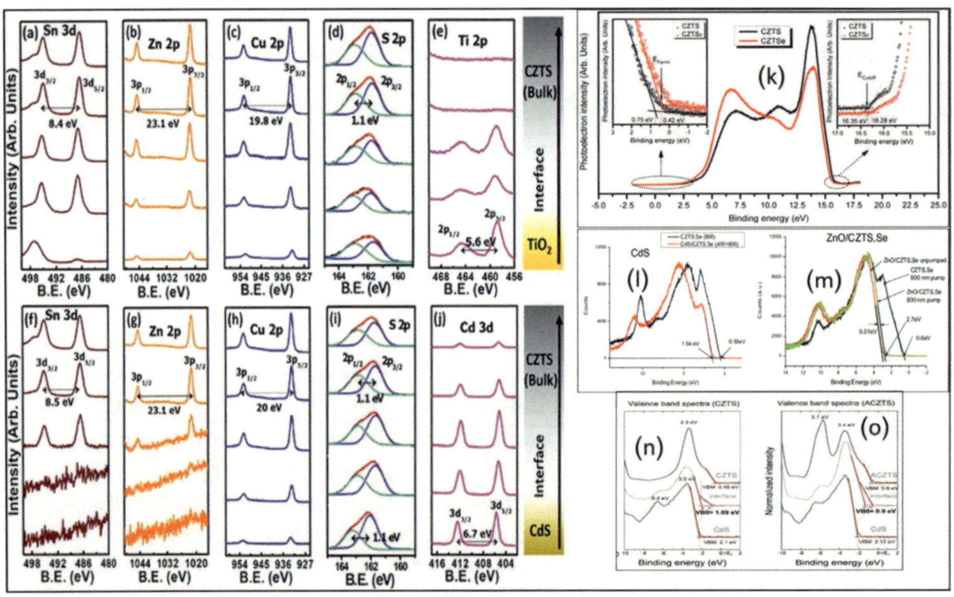

Figure 8.
High-resolution core-level XPS spectra of the CZTS-TiO2 interface (a) Sn 3d, (b) Zn 3p, (c) Cu 2p, (d) S 2p, (e) Ti 2p, and of the CZTS-CdS interface (f) Sn 3d, (g) Zn 2p, (h) Cu 2p, (i) S 2p, and (j) Cd 3d. XPS spectra are recorded in the direction from the top surface of the buffer layer to the CZTS (bulk). (Ref. [55]) (k) The typical UPS spectra of CZTS and CZTSe obtained at excitation photon energy ~21.2 eV. On the left side, inset shows the linear extrapolation of the UPS spectra at low binding energy for the estimation of EFermi, whereas on the right side, inset shows the linear extrapolation of the UPS spectra at high binding energy for the estimation of Ecutof (Ref. [56]), (l) UPS spectra on CdS/CZTS,Se and bare CZTS,Se comparison under flat-band conditions. (m) UPS spectra on ZnO/CZTS,Se (pumped and unpumped) compared with bare CZTS,Se under flat-band condition [figures are reproduced from Ref. [57]]. (n), (o) UPS spectra of CdS, interface, and CZTS/ACZTS (silver-mixed CZTS). The term VBM denotes the valence band maximum position with respect to the Fermi level (EF), while the valence band offset (VBO) was determined from the difference in the VBM positions at the interface [Ref. [58]].

A critical application of UPS in heterojunction-based devices, such as CZTS solar cells, involves determining the conduction band offset (CBO) and valence band offset (VBO), which are key parameters in the interface analysis. By determining the VBO and CBO, researchers can assess energy level alignment at interfaces, a factor crucial for optimizing device performance. The VBO can be calculated using the formula:

$$VBO = E_{VB}^{b} - E_{VB}^{a} - V_{bb} \tag{1}$$

where E_{VB}^{b} and E_{VB}^{a} denote the valence band edges of the buffer layer and absorber layer, respectively, and V_{bb} is the band bending at the interface [59]. This band bending, derived from core-level shifts of elements in the buffer and absorber layers, provides insights into the interface's electronic behavior and can signify favorable or detrimental conditions for electron transfer. To calculate V_{bb} the expression is:

$$V_{bb} = \left(E_{CL}^{a} - E_{CL}^{a}(i)\right) - (E_{CL}^{b}(i) - E_{CL}^{b}) \tag{2}$$

where E_{CL}^{a} and E_{CL}^{b} values correspond to core-level energies within each bulk and at the interface. A negative VBO indicates the buffer layer's valence band edge is positioned lower than that of CZTS, which influences the electronic properties at the heterojunction. UPS also measures the valence band maximum (VBM) for materials like CZTS, TiO_2, and CdS, providing insights into heterojunction behavior. Using linear extrapolation, the VBM positions (relative to the Fermi level) for CZTS, CdS, and TiO_2 are found to be approximately −0.62, −1.72, and − 3.0 eV, respectively [55]. UPS measurements reveal that the average band bending at the CZTS-CdS and CZTS-TiO_2 interfaces is around 0.08 and 0.27 eV, respectively, with valence band offsets of approximately −1.02 eV for CZTS-CdS and − 2.1 eV for CZTS-TiO_2. The CBO, a critical factor for efficient electron transfer and minimal recombination losses, can be calculated with:

$$CBO = E_{g}^{b} - E_{g}^{a} - VBO \tag{3}$$

where E_{VB}^{b} and E_{VB}^{a} values represent the buffer and absorber layers' bandgaps. These measurements show that the CZTS-TiO_2 interface has a favorable "spike-like" configuration within the optimal CBO range (0–0.4 eV), enhancing photovoltaic efficiency. The bandgaps of CZTS, TiO_2, and CdS are estimated at 1.54, 3.8, and 2.4 eV, respectively. The calculated CBO at the CZTS-CdS interface is −0.16 eV, while at CZTS-TiO_2, it is +0.17 eV, which aligns with achieving high efficiency in CZTS-based solar cells [55]. Additionally, UPS helps examine surface modifications through processes like potassium cyanide (KCN) etching, which alters the CZTS surface composition and band structure. Following KCN etching, the surface bandgap for CZTS increases from 1.53 to 1.92 eV, potentially due to the removal of surface contaminants [60]. This modification also raises the recombination barrier, resulting in a 40% efficiency increase in KCN-CZTS solar cells, with improved open-circuit voltage (V_{oc}) attributed to a favorable CBO [61].

Further UPS and XPS analyses have revealed that, at high-temperature substrate bonding (TSB), the interface maintains stability, indicating minimized charge accumulation and potential fluctuations. In contrast, low TSB processes exhibit significant VBM shifts, suggesting the formation of interfacial dipoles and potential electrostatic fluctuations that can degrade performance by increasing recombination losses and decreasing V_{oc} [57, 58].

4. CZTS-based solar cell devices

4.1 CZTS device fabrication

The fabrication of CZTS solar cells typically follows a structured configuration, which includes soda-lime glass (SLG)/Mo/Al2O3/CZTS/CdS/intrinsic ZnO (i-ZnO)/ITO/Al. The aluminum oxide (Al_2O_3) layer is deposited using magnetron sputtering on a Mo-coated SLG. CZTS thin films are subsequently deposited on the Al2O3-covered Mo/SLG substrates through co-sputtering of Cu, ZnS, and SnS (tin sulfide) in the same system [36]. The precursors have average compositions of Cu/(Zn + Sn) and Zn/Sn ratios of 0.7–0.8 and approximately 1.2, respectively. CZTS absorbers, with a thickness of around 800 nm, are prepared through a sulfurization process in an S and SnS atmosphere at a peak temperature of 560°C for 3 minutes, utilizing ramping and cooling rates of 10 and 30°C/min, respectively. The samples are preserved in an N_2 atmosphere to prevent oxidation before the application of the moisture-assisted post-deposition annealing (MAPDA) process on selected samples [62]. The deposition process involved varying the duration of CdS layer formation while maintaining a constant deposition time for the ZnS layer. This strategic variation enabled fine-tuning of the bilayer's optical and electronic properties, optimizing its function as an interface buffer. Comparative analysis demonstrated a notable efficiency enhancement; devices with a traditional CdS buffer layer achieved an efficiency of 7.2%, whereas incorporating a thin ZnS film as a co-buffer boosted efficiency to 10%. This improvement highlights the potential of CdS/ZnS bilayer buffer systems in enhancing CZTSe device performance by facilitating better light absorption and improved charge carrier dynamics at the interface, positioning it as a promising approach for next-generation photovoltaic technology [63]. Further details of the CZTS device fabrication methods can be found in prior publications.

4.2 Influence of pre-treatment on device performance

The pre-treatment conditions of CZTS films significantly affect the optoelectronic properties of CZTSSe absorber materials and the photovoltaic performance of CZTSSe solar cells [64]. The structure and cross-section SEM of the investigated CZTSSe solar cells, along with energy level diagrams, are detailed in **Figure 9(a, b)**. The current-voltage (I-V) characteristics of the best-performing CZTSSe devices, shown in **Figure 9(c)**, were obtained with films sulfurized at 420°C. These devices exhibited a power conversion efficiency (PCE) of 3.48%, with the highest open circuit voltage (V_{oc} = 300 mV) and fill factor (FF = 39.4%) among all devices tested. In contrast, the lowest PCEs were observed in CZTSSe solar cells made from CZTS films that were either not sulfurized or sulfurized at a higher temperature of 450°C [64]. The photovoltaic parameters show a clear trend correlating device performance with the pre-treatment conditions of the CZTS films. The increase in PCE is attributed primarily to improvements in the fill factor and open-circuit voltage, which correlate well with structural analysis results indicating a lower degree of structural disorder and improved crystallinity in films sulfurized at 420°C.

4.3 Influence of sulfurization temperature and device architecture

Sulfurization of CZTS films is crucial for achieving high-performance CZTSSe solar cells. However, even minor variations in annealing conditions can substantially impact the quality of CZTSSe films and the resultant device performance. The

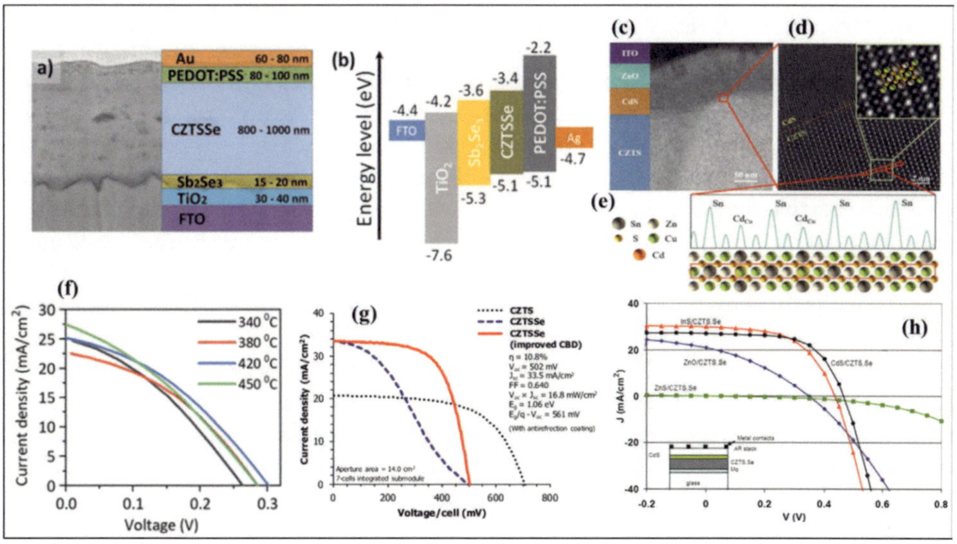

Figure 9.
(a) Structure and characteristics of CZTSSe solar cells. Focused ion beam (FIB)-made cross-section SEM (left) and schematic CZTSSe solar device structure (right). Energy level diagram of the materials used in CZTSSe solar cell device fabrication (Ref. [64]). (c) Bright-field TEM image of cross-sectional CZTS device using the moisture-assisted post-deposition annealing (MAPDA) treatment, with the layer identification shown schematically at left. d) Filtered atomic resolution high-angle annular dark-field (HAADF) image of a region near the CdS/CZTS heterointerface. e) The intensity profile of a row of cations marked by the red rectangle shows the cation exchange (Cd occupying Cu sites), concerning the kesterite structure depicted in the bottom schema. (Ref. [62]) (f) Current-voltage curves of the CZTSSe devices prepared from CZTS films annealed at different temperatures. (Ref. [64]) (g) A new record J-V curve for a CZTSSe submodule having 10.8% efficiency (red solid curve). For comparison, J-V curves for CZTS and CZTSSe are shown in black dotted and blue dashed curves, respectively. (h) J-V curves for the ZnO, ZnS, In_2S_3, and CdS buffers (CZTS,Se-based device). Inset shows device schematic (Ref. [57]).

introduction of an antimony selenide (Sb_2Se_3) buffer layer in device fabrication is another critical factor. The distribution of photovoltaic parameters in CZTSSe solar cells with a Sb_2Se_3 buffer layer exhibits a lower standard deviation compared to those fabricated with a CdS buffer layer, which is attributed to the favorable energy level alignment between Sb_2Se_3 and CZTSSe and the greater stability of the Sb_2Se_3 layer [65]. The incident photon-to-electron conversion efficiency (IPCE) of CZTSSe solar cells demonstrates a strong correlation with short-circuit current density (J_{SC}) values. The highest conversion efficiencies were observed in cells made from films sulfurized at 420 and 450°C, particularly in the longer wavelength region. This indicates that sulfurization at these temperatures enhances the material quality necessary for efficient solar cell operation, likely due to improved charge extraction and collection efficiencies across the spectral range [66]. The integrated current density from the IPCE spectra also shows a strong correlation with the J_{SC} values obtained from current-voltage measurements.

Copper zinc tin sulfide (CZTS) is emerging as a non-toxic alternative to copper indium gallium selenide (CIGS) in photovoltaic applications due to its composition of earth-abundant elements. However, CZTS-based absorbers face significant challenges, including a high density of charged defects and the formation of secondary phases, both of which hinder device efficiency. The high density of defects, particularly Cu and Zn antisite defects, leads to bandgap tailing and reduced minority carrier lifetime. This defect-related bandgap tailing contributes to an open-circuit voltage (V_{oc}) deficit, impairing overall device performance. Additionally, the narrow chemical stability region of CZTS necessitates precise stoichiometric control at the nanoscale to avoid the formation of secondary phases, which can further reduce current collection efficiency. To mitigate these issues, post-annealing treatments at low temperatures have been employed to reduce defect density. For instance, Wang et al. [67] reported a conversion efficiency of 12.6% using such techniques. Zhou et al. [13] achieved a certified total area efficiency of 13.8% by optimizing processing conditions, particularly through controlling pressure at elevated temperatures, resulting in fewer defects and enhanced crystallinity. Despite these advancements, controlling secondary phases remains challenging, as CZTS typically requires off-stoichiometric, Cu-poor, and Zn-rich compositions to achieve higher efficiencies. The presence of secondary phases with bandgaps smaller than that of the absorber can introduce electronic states within the absorber's bandgap, acting as traps and recombination centers, further exacerbating V_{oc} losses. Conversely, secondary phases with larger bandgaps can create narrow conduction pathways, increasing series resistance (Rs) and lowering the fill factor (FF) of the solar cell.

To address these issues, researchers are exploring the partial substitution of absorber constituents with elements like selenium (Se), germanium (Ge), sodium (Na), and antimony (Sb). Furthermore, the interface between the absorber and buffer layer, particularly the suboptimal band alignment at the CZTS/CdS interface, poses additional challenges. The cliff-like band alignment reduces the potential difference between the quasi-Fermi levels of the p-type CZTS and the n-type CdS, limiting device performance. Alternative buffer layers, such as indium sulfide (In_2S_3) and zinc sulfide (ZnS), are being investigated to improve this alignment. Additionally, the use of molybdenum (Mo) as a back contact material is problematic due to its tendency to react with the absorber, forming Mo chalcogenides and decomposition products that introduce a Schottky barrier, increase Rs, and reduce FF. Barrier layers, such as titanium nitride (TiN), zinc oxide (ZnO), and titanium diboride (TiB2), are being explored to mitigate the formation of Mo chalcogenides. Alternative back contact materials like gold (Au), tungsten (W), tantalum (Ta), niobium (Nb), fluorine-doped tin oxide (FTO), and indium tin oxide (ITO) are also under investigation.

5. Conclusion and future prospects

Recent advancements in kesterite-related materials, particularly copper zinc tin sulfide (CZTS), highlight significant potential in merging both the electrical and optical properties of low-cost, abundant, and readily available elements. These developments suggest the feasibility of producing high-efficiency photovoltaic (PV) devices using materials with minimal environmental impact. A variety of deposition techniques, including both vacuum and non-vacuum-based approaches, have been investigated for fabricating CZTS thin films. The current progress in developing pure CZTS thin-film solar cells has achieved a maximum efficiency of 8.4%. However, this efficiency remains lower than those of copper indium selenide (CIS) and copper indium gallium selenide (CIGS) thin-film solar cells, which are already commercially viable and exhibit conversion efficiencies exceeding 15%. To bridge this gap, a deeper understanding of the defect nature within the absorber layer, as well as the design and optimization of intra- and interlayer interfaces, is crucial. Success in this area will require comprehensive material characterizations and innovative device engineering. Additionally, electrical modeling, which has significantly benefited from the experience gained in copper indium gallium sulfur selenide (CIGSSe) technology development, is essential for optimizing solar cell device fabrication. Models that explain recombination mechanisms affecting performance, though partially informed by CIGSSe models, still lack the comprehensiveness needed to cover the full range of experimental observations in CZTS. Developing such a model is critical for identifying performance-limiting mechanisms and optimizing the buffer, window, and back contact layers in line with the electronic and optical properties of the kesterite absorber.

Furthermore, the sulfurization temperature of CZTS thin films plays a pivotal role in determining the quality of CZTSSe absorbing material and the overall performance of CZTSSe-based solar cells. An additional low-temperature sulfurization step is necessary to produce high-quality CZTSSe absorber materials. The choice of sulfurization temperature significantly influences the final quality of CZTSSe films and the solar cells' performance. Therefore, optimizing sulfurization conditions and selecting an appropriate buffer layer are crucial for enhancing the quality and efficiency of CZTSSe solar cell devices.

Author details

Sadhu K. Kolekar[1*], Rahul Ahir[2], Kiran Diwate[3] and Haribhahu Borate[1]

1 PDEA's, Annasaheb Waghire College of Science, Arts and Commerce, Pune, India

2 Department of Physics, Dr. Vishwanath Karad MIT World Peace University, Pune, India

3 Department of Physics, School of Basic and Applied Sciences, JSPM University, Pune, India

*Address all correspondence to: kolekarsadhu@gmail.com

References

[1] Maka AOM, Alabid JM. Solar energy technology and its roles in sustainable development. Clean Energy. 2022;**6**(3):476-483. DOI: 10.1093/ce/zkac023

[2] Ou Q et al. Strong depletion in hybrid perovskite p–n junctions induced by local electronic doping. Advanced Materials. 2018;**30**(15):1705792-1705802. DOI: 10.1002/adma.201705792

[3] Mishra A, Sarbapalli D, Rodríguez O, Rodríguez-López J. Electrochemical imaging of interfaces in energy storage via scanning probe methods: Techniques, applications, and prospects. Annual Review of Analytical Chemistry. 2023;**16**(1):93-115. DOI: 10.1146/annurev-anchem-091422-110703

[4] Jha MK, Subramaniam C. Design principles for manipulating electrochemical interfaces in solid-state supercapacitors for wearable applications. ACS Omega. 2021;**6**(12):7970-7978. DOI: 10.1021/acsomega.1c00172

[5] Wang W, Zhang D, Liu R, Gangadharan DT, Tan F, Saidaminov MI. Characterization of interfaces: Lessons from the past for the future of perovskite solar cells. Journal of Semiconductors. 2022;**43**(5):051202. DOI: 10.1088/1674-4926/43/5/051202

[6] Haddad M, Belhadi A, Boudjellal L, Trari M. Photocatalytic hydrogen production on the hetero-junction CuO/ZnO. International Journal of Hydrogen Energy. 2021;**46**(75):37556-37563. DOI: 10.1016/j.ijhydene.2020.11.053

[7] Cherouana A, Labbani R. Study of CZTS and CZTSSe solar cells for buffer layers selection. Applied Surface Science. 2017;**424**:251-255. DOI: 10.1016/j.apsusc.2017.05.027

[8] Dzade NY. First-principles insights into the electronic structure, optical and band alignment properties of earth-abundant Cu2SrSnS4 solar absorber. Scientific Reports. 2021;**11**(1):4755. DOI: 10.1038/s41598-021-84037-8

[9] Haddout A, Fahoume M, Qachaou A, Raidou A, Lharch M. Understanding effects of defects in bulk Cu2ZnSnS4 absorber layer of kesterite solar cells. Solar Energy. 2020;**211**:301-311. DOI: 10.1016/j.solener.2020.09.067

[10] Korade SD et al. Improving the device performance of CZTSSe thin-film solar cells via indium doping. ACS Applied Materials & Interfaces. 2023. pp. 57183-57191. DOI: 10.1021/acsami.3c13813

[11] Nakamura M, Yamaguchi K, Kimoto Y, Yasaki Y, Kato T, Sugimoto H. Cd-free Cu(In,Ga)(Se,S) 2 thin-film solar cell with record efficiency of 23.35%. IEEE Journal of Photovoltaics. 2019;**9**(6):1863-1867. DOI: 10.1109/JPHOTOV.2019.2937218

[12] Kim S, Park J-S, Hood SN, Walsh A. Lone-pair effect on carrier capture in Cu 2 ZnSnS 4 solar cells. Journal of Materials Chemistry A. 2019;**7**(6):2686-2693. DOI: 10.1039/C8TA10130B

[13] Zhou J et al. Control of the phase evolution of kesterite by tuning of the selenium partial pressure for solar cells with 13.8% certified efficiency. Nature Energy. 2023;**8**(5):526-535. DOI: 10.1038/s41560-023-01251-6

[14] Aka AH, Bouich A, Aka B, Soucase BM. Synthesis and

characterization of kesterite Cu2ZnSn(SxSe1-x)4 thin films with low-cost for efficient solar cells. Results in Optics. 2023;**13**:100507. DOI: 10.1016/j.rio.2023.100507

[15] Mitzi DB, Gunawan O, Todorov TK, Wang K, Guha S. The path towards a high-performance solution-processed kesterite solar cell. Solar Energy Materials & Solar Cells. 2011;**95**(6):1421-1436. DOI: 10.1016/j.solmat.2010.11.028

[16] Hood SN et al. Status of materials and device modelling for kesterite solar cells. Journal of Physics: Energy. 2019;**1**(4):042004. DOI: 10.1088/2515-7655/ab2dda

[17] Kim JY, Lee J-W, Jung HS, Shin H, Park N-G. High-efficiency perovskite solar cells. Chemical Reviews. 2020;**120**(15):7867-7918. DOI: 10.1021/acs.chemrev.0c00107

[18] Tikkiwal VA, Kwatra P, Singh SV, Chandola D, Gandhoke H. A review on sputter deposited CZTS based thin film solar cells. In: 2018 International Conference on Computing, Power and Communication Technologies (GUCON). Greater Noida, India: IEEE; 2018. pp. 400-404. DOI: 10.1109/GUCON.2018.8674967

[19] Demir KÇ. The investigation of the corrosion behavior of CZTS thin films prepared via electrodeposition. Materials Science in Semiconductor Processing. 2021;**123**:105553. DOI: 10.1016/j.mssp.2020.105553

[20] Qiu L, Xu J, Cai W, Xie Z, Yang Y. Fabrication of Cu2ZnSnS4 thin films by microwave assisted sol-gel method. Superlattices and Microstructures. 2019;**126**:83-88. DOI: 10.1016/j.spmi.2018.12.020

[21] Vanalakar SA et al. A review on pulsed laser deposited CZTS thin films for solar cell applications. Journal of Alloys and Compounds. 2015;**619**:109-121. DOI: 10.1016/j.jallcom.2014.09.018

[22] Bhosale SM, Suryawanshi MP, Gaikwad MA, Bhosale PN, Kim JH, Moholkar AV. Influence of growth temperatures on the properties of photoactive CZTS thin films using a spray pyrolysis technique. Materials Letters. 2014;**129**:153-155. DOI: 10.1016/j.matlet.2014.04.131

[23] Khushaim M, Alamri S, Kattan N, Jaber A, Alamri S. Study of kesterite Cu2ZnSnS4 (CZTS) thin films deposited by spray technique for photovoltaic applications. Journal of Taibah University for Science. 2021;**15**(1):329-339. DOI: 10.1080/16583655.2021.1978809

[24] Redinger A, Berg DM, Dale PJ, Siebentritt S. The consequences of Kesterite equilibria for efficient solar cells. Journal of the American Chemical Society. 2011;**133**(10):3320-3323. DOI: 10.1021/ja111713g

[25] Nguyen TH, Harada T, Nakanishi S, Ikeda S. Cu2ZnSnS4-based thin film solar cells with more than 8% conversion efficiency obtained by using a spray pyrolysis technique. In: 2016 IEEE 43rd Photovoltaic Specialists Conference (PVSC). Portland, OR, USA: IEEE; 2016. pp. 0470-0472. DOI: 10.1109/PVSC.2016.7749637

[26] Ben Arbia M, Helal H, Comini E. Recent advances in low-dimensional metal oxides via sol-gel method for gas detection. Nanomaterials. 2024;**14**(4):359. DOI: 10.3390/nano14040359

[27] Yang X et al. Thiourea with sulfur-donor as an effective additive for enhanced performance of lead-free double perovskite photovoltaic cells. Journal of Colloid and Interface Science.

2022;**628**:476-485. DOI: 10.1016/j.jcis.2022.07.165

[28] Courel M, Andrade-Arvizu JA, Guillén-Cervantes A, Nicolás-Marín MM, Pulgarín-Agudelo FA, Vigil-Galán O. Optimization of physical properties of spray-deposited Cu2ZnSnS4 thin films for solar cell applications. Materials and Design. 2017;**114**:515-520. DOI: 10.1016/j.matdes.2016.10.068

[29] Suryawanshi MP et al. CZTS based thin film solar cells: A status review. Materials and Technologies. 2013;**28**(1-2):98-109. DOI: 10.1179/1753555712Y.0000000038

[30] Palve BM, Jadkar SR, Pathan HM. A simple chemical route to synthesize the umangite phase of copper selenide (Cu3Se2) thin film at room temperature. Journal of Semiconductors. 2017;**38**(6):063003. DOI: 10.1088/1674-4926/38/6/063003

[31] Kamoun N, Bouzouita H, Rezig B. Fabrication and characterization of Cu2ZnSnS4 thin films deposited by spray pyrolysis technique. Thin Solid Films. 2007;**515**(15):5949-5952. DOI: 10.1016/j.tsf.2006.12.144

[32] Kishore Kumar YB, Suresh Babu G, Uday Bhaskar P, Sundara Raja V. Preparation and characterization of spray-deposited Cu2ZnSnS4 thin films. Solar Energy Materials & Solar Cells. 2009;**93**(8):1230-1237. DOI: 10.1016/j.solmat.2009.01.011

[33] Rajeshmon VG, Kartha CS, Vijayakumar KP, Sanjeeviraja C, Abe T, Kashiwaba Y. Role of precursor solution in controlling the opto-electronic properties of spray pyrolysed Cu2ZnSnS4 thin films. Solar Energy. 2011;**85**(2):249-255. DOI: 10.1016/j.solener.2010.12.005

[34] Patil PS. Versatility of chemical spray pyrolysis technique. Materials Chemistry and Physics. 1999;**59**(3):185-198. DOI: 10.1016/S0254-0584(99)00049-8

[35] Johnson DR. Microstructure of electrodeposited CdS/CdTe cells. Thin Solid Films. 2000;**361-362**:321-326. DOI: 10.1016/S0040-6090(99)00779-8

[36] Cabas-Vidani A et al. Influence of the rear Interface on composition and photoluminescence yield of CZTSSe absorbers: A case for an Al_2O_3 intermediate layer. ACS Applied Materials & Interfaces. 2021;**13**(16):19487-19496. DOI: 10.1021/acsami.1c02437

[37] Araki H et al. Preparation of Cu2ZnSnS4 thin films by sulfurizing electroplated precursors. Solar Energy Materials & Solar Cells. 2009;**93**(6-7):996-999. DOI: 10.1016/j.solmat.2008.11.045

[38] Tanaka K, Moritake N, Uchiki H. Preparation of Cu2ZnSnS4 thin films by sulfurizing sol–gel deposited precursors. Solar Energy Materials & Solar Cells. 2007;**91**(13):1199-1201. DOI: 10.1016/j.solmat.2007.04.012

[39] Tanaka K, Oonuki M, Moritake N, Uchiki H. Cu2ZnSnS4Cu2ZnSnS4 thin film solar cells prepared by non-vacuum processing. Solar Energy Materials & Solar Cells. 2009;**93**(5):583-587. DOI: 10.1016/j.solmat.2008.12.009

[40] Moritake N, Fukui Y, Oonuki M, Tanaka K, Uchiki H. Preparation of Cu 2 ZnSnS 4 thin film solar cells under non-vacuum condition. Physica Status Solidi C: Current Topics in Solid State Physics. 2009;**6**(5):1233-1236. DOI: 10.1002/pssc.200881158

[41] Gudmundsson JT. Physics and technology of magnetron sputtering discharges. Plasma Sources Science

and Technology. 2020;**29**(11):113001. DOI: 10.1088/1361-6595/abb7bd

[42] Efaz ET et al. A review of primary technologies of thin-film solar cells. Engineering Research Express. 2021;**3**(3):032001. DOI: 10.1088/2631-8695/ac2353

[43] Yang K-J et al. Flexible Cu2ZnSn(S,Se)4 solar cells with over 10% efficiency and methods of enlarging the cell area. Nature Communications. 2019;**10**(1):2959. DOI: 10.1038/s41467-019-10890-x

[44] Kaur K, Ghosh A, Nisika N, Kumar M. Strain modulation for enhancing Cu–Zn ordering in CZTS absorber layer using seed layer assisted growth for efficient carrier transport. Applied Physics Letters. 2021;**118**(25):252108-252116. DOI: 10.1063/5.0055736

[45] Nagoya A, Asahi R, Wahl R, Kresse G. Defect formation and phase stability of Cu_2ZnSnS_4 photovoltaic material. Physical Review B. 2010;**81**(11):113202. DOI: 10.1103/PhysRevB.81.113202

[46] Valentini M et al. Effect of the order-disorder transition on the optical properties of Cu2ZnSnS4. Applied Physics Letters. 2016;**108**(21):211909-211914. DOI: 10.1063/1.4952973

[47] Aldalbahi A, Mkawi EM, Ibrahim K, Farrukh MA. Effect of sulfurization time on the properties of copper zinc tin sulfide thin films grown by electrochemical deposition. Scientific Reports. 2016;**6**(1):32431. DOI: 10.1038/srep32431

[48] Villanueva YY, Liu D-R, Cheng PT. Pulsed laser deposition of zinc oxide. Thin Solid Films. 2006;**501**(1-2):366-369. DOI: 10.1016/j.tsf.2005.07.152

[49] Katagiri H et al. Development of CZTS-based thin film solar cells. Thin Solid Films. 2009;**517**(7):2455-2460. DOI: 10.1016/j.tsf.2008.11.002

[50] Moriya K, Tanaka K, Uchiki H. Fabrication of Cu2ZnSnS4 thin-film solar cell prepared by pulsed laser deposition. Japanese Journal of Applied Physics. 2007;**46**(9R):5780. DOI: 10.1143/JJAP.46.5780

[51] Moriya K, Tanaka K, Uchiki H. Cu2ZnSnS4 thin films annealed in H2S atmosphere for solar cell absorber prepared by pulsed laser deposition. Japanese Journal of Applied Physics. 2008;**47**(1S):602. DOI: 10.1143/JJAP.47.602

[52] Hossain MA et al. Atomic layer deposition enabling higher efficiency solar cells: A review. Nano Materials Science. 2020;**2**(3):204-226. DOI: 10.1016/j.nanoms.2019.10.001

[53] Giraldo S et al. Study and optimization of alternative MBE-deposited metallic precursors for highly efficient kesterite CZTSe:Ge solar cells. Progress in Photovoltaics: Research and Applications. 2019;**27**(9):779-788. DOI: 10.1002/pip.3147

[54] Nakayama Y, Tsuruta R, Koganezawa T. 'Molecular beam epitaxy' on organic semiconductor single crystals: Characterization of well-defined molecular interfaces by synchrotron radiation X-ray diffraction techniques. Materials (Basel). 2022;**15**(20):7119. DOI: 10.3390/ma15207119

[55] Nisika N, et al. Energy level alignment and nanoscale investigation of a-TiO2/Cu-Zn-Sn-S interface for alternative electron transport layer in earth abundant Cu-Zn-Sn-S solar cells. Journal of Applied Physics. 2019;**126**(19):193104-193116. DOI: 10.1063/1.5121874

[56] Pandharkar S et al. Probing the effect of selenium substitution in kesterite-Cu2ZnSnS4 nanocrystals prepared by hot injection method. Journal of Materials Science: Materials in Electronics. 2019;**30**(15):14781-14790. DOI: 10.1007/s10854-019-01851-7

[57] Haight R et al. CdS and Cd-free buffer layers on solution phase grown Cu 2 ZnSn(S x Se 1- x) 4 :Band alignments and electronic structure determined with femtosecond ultraviolet photoelectron spectroscopy. MRS Proceedings. 2014;**1638**:1-6 mrsf13-1638-w10-02-z04-02. DOI: 10.1557/opl.2014.196

[58] Gansukh M et al. Energy band alignment at the heterointerface between CdS and Ag-alloyed CZTS. Scientific Reports. 2020;**10**(1):18388. DOI: 10.1038/s41598-020-73828-0

[59] Kraut EA, Grant RW, Waldrop JR, Kowalczyk SP. Precise determination of the valence-band edge in X-ray photoemission spectra: Application to measurement of semiconductor interface potentials. Physical Review Letters. 1980;**44**(24):1620-1623. DOI: 10.1103/PHYSREVLETT.44.1620

[60] Bär M et al. Impact of KCN etching on the chemical and electronic surface structure of Cu2ZnSnS4 thin-film solar cell absorbers. Applied Physics Letters. 2011;**99**(15):152111-152114. DOI: 10.1063/1.3650717

[61] Jagadish K et al. Interface-centric strategies in Kesterite solar cells: Addressing challenges, solutions, and future directions for efficient solar-harvesting technologies. Small. 2024. pp. 2402048-2402072. DOI: 10.1002/smll.202402048

[62] Sun H et al. Defect engineering for efficient Cu 2 ZnSnS 4 solar cells via moisture-assisted post-deposition annealing. Advanced Optical Materials. 2022;**10**(18):2200607-2200616. DOI: 10.1002/adom.202200607

[63] Hernández-Calderón V et al. CdS/ZnS bilayer thin films used As buffer layer in 10%-efficient Cu2ZnSnSe4 solar cells. ACS Applied Energy Materials. 2020;**3**(7):6815-6823. DOI: 10.1021/acsaem.0c00937

[64] Pakštas V et al. Improvement of CZTSSe film quality and superstrate solar cell performance through optimized post-deposition annealing. Scientific Reports. 2022;**12**(1):16170. DOI: 10.1038/s41598-022-20670-1

[65] Tiwari KJ et al. Efficient Sb2Se3/CdS planar heterojunction solar cells in substrate configuration with (hk0) oriented Sb2Se3 thin films. Solar Energy Materials & Solar Cells. 2020;**215**:110603. DOI: 10.1016/j.solmat.2020.110603

[66] Emrani A, Vasekar P, Westgate CR. Effects of sulfurization temperature on CZTS thin film solar cell performances. Solar Energy. 2013;**98**:335-340. DOI: 10.1016/j.solener.2013.09.020

[67] Wang S et al. Smooth perovskite thin films and efficient perovskite solar cells prepared by the hybrid deposition method. Journal of Materials Chemistry A. 2015;**3**(28):14631-14641. DOI: 10.1039/C5TA03593G